HAZARDOUS CHEMICALS

A Manual for Schools and Colleges

**Scottish Schools
Science Equipment Research Centre**

Oliver & Boyd

During the preparation of this manual the Centre has drawn on other works. These include:

Hazards in the Chemical Laboratory edited by G. D. Muir, Royal Institute of Chemistry/Chemical Society, 1971;

Handbook of Reactive Chemical Hazards by L. Bretherick, Butterworth, 1975;

Dangerous Properties of Industrial Materials edited by Irving Sax, Van Nostrand Reinhold, 1975;

The Care, Handling and Disposal of Dangerous Chemicals by P. J. Gaston, Northern Publishers, 1965;

and the *Threshold Limit Values* for 1977, published by the American Conference of Governmental Industrial Hygienists, as presented in Environmental Health Note 15/77 by the Health and Safety Executive.

Use has been made of the experience of many unnamed teachers and our thanks are also due to them.

Oliver & Boyd
Robert Stevenson House
1-3 Baxter's Place
Leith Walk
Edinburgh EH1 3BB

A Division of Longman Group Ltd.

ISBN 0 05 003204 6

First published 1979
Reprinted 1980

Printed in Hong Kong by
Wilture Enterprises (International) Ltd

Foreword

Since it was established in 1964 the Scottish Schools Science Equipment Research Centre has operated on behalf of all of the local education authorities in Scotland to offer a service of constructive and practical advice and example to schools. This it has done to excellent effect as is evidenced by the many comments received both from establishments in this country and overseas.

Particularly with the introduction of the Health and Safety at Work Act, attention has quite rightly been focused on the question of safety in science laboratories. A large number of publications exist which are each good in their field but as with other problems our aim is to find one which is appropriate to the very specific problems which face science staff in our schools. For this reason the Centre has compiled this manual to be an immediate source of reference and guidance, and what follows has been gathered from practical professional contributors presented in a style which should be of daily value to the practising teacher.

Thanks are due to the many individuals and organisations for their contribution to compiling and checking this publication so expertly. One particular person, Mr Hugh Medine, recently retired as Assistant Director of the Centre, bore the special responsibility for preparing much of the initial material, and his considerable skill in producing it will be a particular testimonial to his career with SSSERC.

J McGinley
Chairman

The natural disposition of those concerned to have safe school laboratories has been reinforced by legislation with penalties for neglect or for infringement. The Health and Safety at Work Act 1974 requires the provision of information, instruction and training of employees by the employer and of arrangements which ensure, as far as is reasonably practicable, the safety and absence of risks to health in connection with the use, handling and storage of articles and substances.

The main part of this manual is an alphabetical collection of loose leaves, each page carrying the information relevant to one particular chemical. Prior consultation of a particular page can make the teacher and technician aware of the likely hazards, recommended handling and storage procedures, and of the measures needed to deal with spillages. Thus they can be forewarned and forearmed. In order to facilitate the use of this manual the language has been reduced to a minimum, whilst every attempt has been made to retain all the necessary information. The nomenclature recommended by the ASE has been used and chemicals are placed alphabetically according to this. However reference has also been made to alternative or traditional names and an alphabetical list of these appears after this introduction.

This manual was drawn up by a large group which included several practising teachers, HM Inspectors and members of the Centre. The substances listed were chosen with a school syllabus in mind, and the manual contains many chemicals not found in other works which cater for more advanced laboratories. Being in loose-leaf form, new pages for other chemicals can be added or, in the event of new information about the hazards associated with a particular chemical becoming available, substitution by a new sheet is possible.

It is hoped that the users of the manual will feed back information, suggest new additions, etc. so that on the basis of experience in school laboratories the manual can evolve and hence increase its usefulness.

Since many chemicals previously assumed to be safe have now been shown to be hazardous, it is sensible to handle all chemicals carefully. Each page lists the nature of the hazard of a particular chemical, but a number of generalisations may be made.

(a) The hazard presented by a mixture of some chemicals is often much greater than that of any alone. For example, a mixture of methanal vapour and hydrogen chloride produces chloromethoxychloromethane which is a very potent carcinogen. Another example is that of a mixture of a strong oxidising agent and a reducing agent such as thermit mixture.

(b) The action of many organic solvents on the skin leads to a defatting of tissues and possibly to dermatitis. A toxic chemical will penetrate the skin more easily if dissolved in such a solvent. For this reason all care should be taken to avoid contact of skin with solvents, and in the event of contamination the skin should be well washed with soap and water.

(c) A less toxic substitute should always be sought. For example, in change of state experiments safer alternatives to naphthalene are some higher aliphatic acids and alcohols, or 1,1,1-trichloroethane used in place of 1,1,2-trichloroethane.

(d) Good practice and technique obviously reduce risk and the extent of contamination. Suitable protective clothing and safety equipment should be used when necessary. *No chemicals should be pipetted by mouth.* Cheap pipette fillers can be made from disposable plastic syringes (used ones should *not* be obtained from hospitals, etc.) and several different types of filler are commercially available.

(e) Good ventilation prevents a build-up of vapour concentrations and is one of the greatest aids to safety in a chemical laboratory.

(f) *Fire.* Each laboratory, preparation room and storeroom should have sited near the exit door:
  (*i*) a bucket of dry sand and a scoop;
  (*ii*) a compressed carbon dioxide or dry powder extinguisher;
  (*iii*) a fibreglass fire blanket.

The fire extinguisher and blanket should be mounted in a readily visible and unobstructed place not more than one metre above the ground. Many laboratories presently have foam extinguishers which are likely to be replaced by the dry powder type. Carbon dioxide is suitable for all fires except those of alkali metals, calcium, magnesium, and their hydrides. Fires of these reactive metals can only be dealt with by specially formulated dry powders, or if small they should be allowed to burn out. Carbon dioxide has a fairly quick knock-down effect, but often lacks effectiveness as it is easily dispersed. On the other hand, a high concentration of carbon dioxide in a small space may asphyxiate the operator. A fire blanket is good for smothering vessels and their contents. A person whose clothes have ignited should be rolled up in it, but it should be realised that the good insulation properties of a blanket may 'cook' him, or that if his head is partially enveloped he may even be gassed.

It cannot be stressed too strongly that in the field of fire fighting both teachers and technicians lack professional training and equipment, and *they should attempt nothing more than first-aid fire fighting of small fires.* A 10 lb carbon dioxide extinguisher in the hands of a trained operator is barely sufficient to extinguish a Winchester of a flammable solvent. The decision to tackle a fire can only be taken by the teacher on the spot, and in the case of a large fire the room(s) should be evacuated, doors closed and the fire authorities summoned.

The fire risk can be reduced by:
  (*i*) storing oxidising and reducing agents separately – none of the former should ever be placed in a waste bin;
  (*ii*) keeping to a minimum the quantities of flammables in the laboratory;
  (*iii*) storage of solvents in a cool store with spark-proof light and switches;
  (*iv*) regularly checking Bunsen burner tubing, replacing where needed and ensuring that it is firmly attached to gas taps;
  (*v*) keeping low the level of flammable vapours and not allowing a build-up to explosive levels. This can be avoided by using a fume cupboard or by good ventilation. The concentrations of many vapours needed to form explosive mixtures with air are often very low. For example, the percentages by volume of ethanol, propanone, carbon disulphide and methane to form an explosive mixture with air are 3.3, 2.6, 1.3 and 5.3 respectively;
  (*vi*) rejection of any cracked vessels and damaged apparatus;
  (*vii*) turning off all sources of ignition during handling of substances with low autoignition temperatures. This is defined as the lowest temperature at which a substance will self-ignite in the absence of a spark or flame. The autoignition temperatures of many organic chemicals are well below the temperatures of flames and sparks, and even a warm hotplate, a radiator or the surface of an electric light bulb can cause ignition of vapours with very low autoignition temperatures, e.g. carbon disulphide, ethoxyethane (diethyl ether);
  (*viii*) the use of waterbaths, heating mantles, hot plates rather than a direct Bunsen flame;
  (*ix*) careful disposal of waste solvent;
  (*x*) carefully washing any area contaminated by a spillage of an oxidising agent.

(g) The Threshold Limit Values have in the latest issue of the relevant Health and Safety Environmental Health Note (EH15/77) been described as the *Threshold Limit Value – Time Weighted Average (TLV–TWA)* and are defined as the time weighted average concentrations to which it is considered that nearly all workers may be repeatedly exposed for an eight-hour day, or 40-hour working week, for a working life without adverse effect.

For certain substances the use of a time weighted average TLV is not appropriate.

Substances whose effects are acute and are fast acting are best controlled by a *ceiling limit* which should not be exceeded even for an instant. These have been designated 'C'.

Some chemicals are rapidly absorbed by the skin (including eyes and mucous membranes by airborne means or by direct contact) to such an extent that appropriate means for prevention of cutaneous absorption is necessary if the TLV is not to be invalidated. Such substances have been designated 'skin'.

A new parameter, the Threshold Limit Value – Short Term Exposure Limit (TLV–STEL), is the maximum concentration to which workers can be exposed for a period of up to fifteen minutes without suffering (*i*) intolerable irritation, (*ii*) chronic or irreversible tissue change, or (*iii*) narcosis of sufficient degree to increase accident proneness, impair self-rescue or materially reduce work efficiency – provided that no more than four excursions per day are permitted, with at least 60 minutes between exposure periods, and provided that the TLV–TWA also is not exceeded. The STEL should also be considered as a maximal allowable concentration or absolute ceiling which is not to be exceeded at any time during the fifteen minute excursion period. For most chemicals the values of STELs lie between one and 1.5 times the TLV–TWA value. With the exception of a few specific instances, e.g. possibly during monitoring of atmosphere following a spillage of mercury, it is extremely unlikely that any measurements of airborne concentrations will be made in a school laboratory. For this reason STEL values have not been included. *Thus the TLV–TWA can be regarded as a time weighted average and the ceiling of a chemical is either equal to it or slightly greater than it.*

The threshold of smell of many chemicals is usually below that of the TLV–TWA values and in many cases smell can act as a warning that the TLV is being approached. Since many gases in dangerous concentrations temporarily impair the sense of smell, a hazard may exist without laboratory workers being aware of it, and the use of smell should be regarded as a useful warning, but not necessarily a reliable one.

In spite of the fact that serious injury is not believed likely as a result of exposure to the TLV concentrations, the *best practice is to maintain concentrations of all atmospheric contaminants as low as is practicable.* These limits are intended for use in the practice of industrial hygiene and should be interpreted and applied only by a person trained in this discipline. They are not intended for use, or for modification in use,

(*i*) as a relative index of hazard or toxicity;

(*ii*) in evaluation or controls of community air pollution nuisances;

(*iii*) in estimating the toxic potential of continuous uninterrupted exposures or other extended work periods;

(*iv*) as proof or disproof of an existing disease or physical condition; or

(*v*) for adoption by countries whose working conditions differ from those of the USA and where substances and processes differ.

TLVs are not sharp dividing lines between 'safe' and 'dangerous' concentrations. Of two substances with similar physiological actions on the body, one with a high TLV and high volatility can clearly present a greater hazard than another which has a low TLV and a high boiling point. Both TLV–TWAs and STELs would seem to be more applicable to both teacher and technician whereas the STEL would seem to be more applicable to most pupils who are in laboratories for only small portions of each week. However it is possible that levels lower than the TLV–TWA should be applied to young persons as they have a lower body mass and possibly a greater sensitivity to certain chemicals than do adults.

Clearly the best methods of reducing the chance that these limits may be exceeded include:

(*i*) good ventilation;

(*ii*) use of fume cupboards;

(*iii*) use of closed flasks with vents or reflux condensers rather than open beakers and flasks;

(*iv*) using small scale preparations especially in cases where a noxious gas is evolved. A simple calculation shows that twenty pupils each burning 0.1 g of sulphur will produce sufficient sulphur dioxide to reach the TLV of 13 mg m^{-3} in a room $(8 \times 8 \times 5)$m^3 if there are no air changes.

(h) Where a chemical is known to cause physical impairment or death following an exposure of several weeks, months or years to a low concentration, this has been described as a chronic effect in the *Hazards* section. Whereas acute effects are usually immediately noticeable by the accompanying irritation, chronic effects are insidious in that little or no warning is apparent, e.g. heavy metals or carcinogens.

(i) Lists of prohibited and controlled carcinogens are available in DES AM 3/70 and SED Circular 825 (1972). According to the latter the manufacture, presence and use of the following chemicals and their salts are *prohibited*:

naphthalen-2-amine (2-naphthylamine)	4-aminobiphenyl
benzidine (biphenyl-4,4'-diamine)	4-nitrobiphenyl
naphthalen-1-amine (1-naphthylamine)	4,4'-dinitrobiphenyl.

While the following chemicals and their salts:

o-tolidine (3,3'-dimethylbiphenyl-4,4'-diamine)
dichlorobenzidine (dichlorobiphenyl-4,4'-diamines)
dianisidine (3,3'-dimethoxybiphenyl-4,4'-diamine)

are *controlled* and should be used in educational establishments only under stringent conditions of control and supervision.

It further states:

'In addition to those listed in The Carcinogenic Substances Regulations 1967 No. 879, a large number of other substances, listed below, have been shown to cause cancerous growths in animals, and it is possible that they could have similar effects in man.

(*i*) Polycyclic aromatics and their derivatives. These include benzpyrene, dimethylbenzanthracene, benzacridine, dibenzcarbazole, tricycloquinazoline.

(*ii*) Aromatic amines, nitro compounds and related compounds. These include 1-nitronaphthalene, 2-nitronaphthalene, 4-aminostilbene, 4-nitrostilbene, 2-acetamidofluorene, o-aminoazotoluene, 4-dimethylaminoazobenzene, and other azo dyes.

(*iii*) N-nitroso compounds of the general formula R.N(NO)R' and the related N-nitrosamides R.N(NO)CO.R'.
These include dimethylnitrosamine, N-nitrosopiperidine, N-methyl-N-nitrosoaniline, N-methyl-N'-nitro-N-nitrosoguanidine, N-methyl-N-nitrosourethane, N-methyl-N-nitrosourea.

(*iv*) Other specific substances such as azoxyethane, 1,2-diethylhydrazine, urethane, thiourea, nitrogen mustard, cyclophosphamide, tretamine, β-propiolactone, myleran, naturally occurring carcinogens such as supinidine, cyasin and the aflotoxins.

(*v*) Complex mixtures such as coal tar and crude petroleum.
This list is by no means exhaustive. The compounds named are indicative of the wide range of substances which may have carcinogenic activity.
It is recommended that:

(*i*) Stocks of these substances in laboratories should be kept to the minimum quantity necessary for educational purposes.

(*ii*) they should be used only under the strict control of the teacher in charge;

(*iii*) the precautions applicable to the substances listed in the carcinogenic substances regulations should be applied to them also.'

Many other chemicals are reported as being either a 'recognised carcinogen' or a 'suspected carcinogen' in the 1975 edition of Sax, but it should be pointed out that there is a very wide range of carcinogenic potency. Many have been so classified for animals and have yet to be shown to cause cancer in humans, whereas others have been shown to be carcinogenic only in very high doses over a long period of time. Wherever reasonable suspicion has been reported it has been recorded in this manual. Many samples of compounds, particularly of certain dyestuffs themselves, may be thought to be relatively safe but can contain carcinogens as impurities. The intention here is not that of prohibiting another group of chemicals but to suggest that it is wise to take extra precautions during the handling of these chemicals.

(j) *Incompatibility.* The accidents so far reported cannot provide a complete list of unsafe combinations of chemicals. A prudent user of this manual should apply 'chemical sense' or intuition to extend the range of this section. For example, sulphuric acid is reported as reacting violently with manganates (VII) (permanganates), chlorates (VII) (perchlorates), and chlorates (V) (chlorates) and it would be wise not to prepare, or at least treat with great caution, any mixture of sulphuric acid and other oxoanions of halogens and transition metals, e.g. chlorate (I) (hypochlorite) or bromate (V), titanate (IV), etc.

(k) The advice given under *Handling* regarding the precautions to be taken (i.e. whether the use of gloves and/or a fume cupboard is necessary) has to cover all situations from that of the teacher or technician who may be frequently handling large amounts of chemicals, often in concentrated form, to that met with in small-scale pupil experiments.

Clearly there are many volatile compounds which may be handled in small quantities in well-ventilated laboratories. A simple calculation of the total amount of volatiles being handled (e.g. 20 pupils each using 2 g of a volatile) and the volume of the room gives the teacher a rough estimate of the maximum possible atmospheric concentration in a laboratory with no air changes. This value is unlikely to be reached as there will be ventilation and since it is most unlikely that all of the volatiles will evaporate. The rough figure obtained gives a teacher an indication as to whether or not the TLV–TWA is likely to be exceeded and to whether use of a fume cupboard will be necessary.

Many dilute solutions may be handled without gloves. Again it is only the teacher on the spot who has knowledge of the experience of his pupils, the particular techniques or processes and scale of the experiment. The sheets indicate the danger due to skin absorption or inhalation and only the teacher or his employer can make a final decision as to whether the use of gloves or a fume cupboard is necessary.

No one material for gloves is resistant to all the chemicals likely to be encountered. Natural rubber is adversely affected by nitric acid, chlorates (I) (hypochlorites), alkanes, arenes and to a lesser extent by halogenated hydrocarbons, esters and some amines. Nitrile rubber, neoprene and PVC have a larger useful working range but the reader is referred to the glove resistance ratings published by James North and Sons (PO Box 3, Hyde, Cheshire SK14 1RL).

(l) *Storage.* The safest system of storage would provide several separate storage areas, one for each group of compatible chemicals and these areas would be dry, cool and well-ventilated. Thus in the event of accidental breakage or fire there is a much reduced chance of a violent reaction. This ideal can seldom be reached, but the philosophy behind it should be used as far as resources permit.

(i) *Poisons.* In the UK the storage of poisonous substances is controlled by the Poisons Rules (1978) and the Health and Safety at Work Act 1974 (in the form of The Packaging and Labelling of Dangerous Substances Regulations 1978). The Poisons Rules require that certain chemicals classified under Schedule 1 be stored in areas to which pupils do not have access. The Packaging and Labelling of Dangerous Substances Regulations require that many chemicals be kept out of the reach of children, that some be kept under lock and key, and that this and other information be printed on the supplier's label. We would recommend that these listed chemicals, and others not listed in schedules but known to be very hazardous, should not be kept in the laboratory but in either a secure storeroom or locked cupboard labelled 'Poisons' which is located in an area not used by pupils. The contents should be regularly checked. There is the possibility that some of the poisons so stored together may be incompatible with each other, e.g. barium peroxide and ethanedioic acid. Considera-

tion should be given to isolating these as much as possible or to having two poisons cupboards. It is recognised that senior pupils under supervision will have some access to poisons and that dilute solutions of reagents, e.g. mercury (II) chloride or barium nitrate, may be made more accessible than the solid reagents.

Teachers in other countries who may not be subject to these rules and classifications should find them a useful guideline.

(ii) *Flammables.* Although the Highly Flammable Liquids and Liquefied Petroleum Gases Regulations 1972 do not apply to schools they can be considered as a guide to good practice. Ideally, flammables should be kept in an outside store or in a storeroom of suitable construction within the building. Electrical switchgear should be outside the store. Up to 50 litres may be stored in a workroom in Winchesters in a cupboard or bin whose fabric is of fire resisting materials. Many metal and some wooden cabinets which comply with the regulations are available. As wood is a better insulating material than steel, a wooden cupboard protects its contents from a rapid rise in temperature (Fire Research Note No. 998 of the Building Research Establishment, Borehamwood). A steel cupboard permits rapid heat transfer from the fire to the contents resulting in breaking of bottles. The distortion of the metal cupboard allows entry of air and escape of hot vapours. It seems reasonable to say that an unvented wooden cupboard offers more protection than a steel one.

(iii) *General store.* Chemicals should be stored in the following groupings at different parts of the store. Containers such as large biscuit tins are useful for enclosing some of the groups.

1. Reducing agents including alkali metals, alkaline earth metals, their hydrides, and other metals especially if powdered. Ideally this group should be in a small second flammables store. Phosphorus should be locked up far from both oxidising and reducing agents.

2. Oxidising agents including oxoacid salts such as nitrites, nitrates, oxoanions of halogens, chlorates (I, III, V and VII) (hypochlorites, chlorites, chlorates and perchlorates), bromates (I, III and V) (hypobromites, bromites, bromates), iodates (I, III, V and VII) (hypoiodites, iodites, iodates and periodates), oxoanions of transition metals, higher oxides and peroxides.

3. Acids. Large Winchesters should be stored at floor level on shallow trays or on PVC sheet or tiles.

4. Alkali solutions should be similarly stored at a different site in the store room.

5. Hydrolysable compounds such as ethanoyl chloride, tin (IV) chloride, silicon tetrachloride and some anhydrides should be kept well away from alkalis and aqueous solutions.

(m) It is useful to keep stored in a small cupboard a 'spillage kit' which contains rubber gloves, leather gauntlets, mop and bucket, non-flammable detergent or emulsifying agent, face visor, sand, sodium hydrogencarbonate, sawdust, and a stiff brush with long handle. Such a kit could for example be kept in a central preparation room serving two or three adjacent laboratories.

Each suite of laboratories should also have stored, either with the spillage kit or with a fire extinguisher, a respirator which is suitable for use with most of the general classes of chemicals. Most makes of activated filter respirators require the use of at least four or five different cartridges each covering a particular range of gases, e.g. for ammonia, acid fumes, organic vapours, metal and metal oxide fumes. Two manufacturers offer suitable general purpose filters fitted to ori-nasal half masks. Sabre Safety Limited[1] supply mask HM/75 and filter B/K which as a general guidance is stated as being able to protect a person against a concentration of 1% by volume of most gases for between 15 and 20 minutes. Canister B/K has a shelf life of three years. Draeger Safety Limited[2] supply the Parat I and Parat II which provide protection against low concentrations of all gases except carbon monoxide for about 10 minutes. These have a shelf life of four years. The use of filter respirators with extremely toxic gases such as carbon monoxide and hydrogen cyanide is not advisable. Sabre supply a full face mask for use in conjunction with their filter. The wearing of a respirator does not enable the teacher to play the role of a professional fire officer; *it only allows him to re-enter a laboratory for the minimum time necessary to dispose of a small spillage and to open windows.* Neither should a filter cartridge respirator be used in atmospheres containing less than 15% by volume of oxygen. Though more expensive, the ideal equipment is a self-contained positive pressure breathing apparatus as, unlike the filter respirator, there is little danger caused if the face mask has not been properly fitted, or if a high concentration of vapour is encountered.

(n) *Disposal and Spillage.* Large amounts of chemicals should be dealt with by specialist contractors. If deposition of a noxious chemical is by a commercial contractor at a tip then the Deposit of Poisonous Waste Act 1972 applies. This requires at least three days' notification of user's intention to be given to the local authorities and water authorities (or purification boards) both in the area in which the material is being tipped and in the area from which it is being removed.

Small quantities are usually best dealt with by a chemistry teacher or competent technician. Volatile water-immiscible solvents can be soaked up on sand, transported to a safe outdoor position and allowed to evaporate, or small amounts can be carefully burned. The transportation of waste solvents to an outdoor site and the burning of them can be very hazardous procedures and should only be carried out by a competent person. Water-soluble solvents such as ethanol, ethanal, methanal and substances such as ethanoic acid and phenylamine after neutralisation can be safely diluted and run to waste with much water.

In schools the quantities of nitrites, nitrates, phosphates and sulphates should cause no problem if greatly diluted and run to waste. Only in areas where lowland river water is used for a water supply will there be a restriction.

In the case of a spillage, solvents can be emulsified by brushing water and a non-flammable dispersing agent into the spillage with a stiff brush. This can be mopped up and washed down the drain with plenty of water.

Trade effluent standards can be obtained from the local Regional Water Authority or from the local Inspector of Drainage and should be written into the appropriate pages of the manual under *Local Conditions.* In the area administered by one water board the total concentration of heavy metals should not exceed 20 mg l^{-1} in aggregate, and in order to reach this acceptable level at the point of entry into a main sewer 20 g of a solid would have to be diluted with a total volume of 1000 litres. A laboratory tap delivering at the rate of 10 litres per minute would need to run for approximately $1\frac{1}{2}$ hours.

(o) *First Aid.* Prevention is better than cure but in spite of good working practice the unforeseeable may still occur. Recommendations given here refer to chemicals listed in this manual, but their general application to other chemicals is not necessarily precluded. After following the simple first aid listed for each chemical, *medical advice should always be sought in cases of laboratory chemicals splashed in* eyes *and in cases where chemicals have entered the* mouth *and possibly been swallowed.*

(i) *Eyes* should first be well irrigated by a *slow* stream of clean water from either a tap or from a special eyewash bottle freshly filled with clean water. Dehydrating agents (e.g. concentrated sulphuric acid or ethanoic anhydride), oxidising agents (e.g. potassium manganate (VII) (potassium permanganate), bromates (I, III and V) (hypobromites, bromites and bromates), chlorates (I, III, V and VII) (hypochlorites, chlorites, chlorates and perchlorates), iodates (I, III, V and VIII) (hypoiodites, iodites, iodates and periodates), peroxides), acids, alkalis, other chemicals which react with moisture to produce alkalis (e.g. calcium dicarbide, sodium), organic solvents and phenols are particularly damaging to eye tissues. In cases which require medical attention, the patient should be removed to hospital with a label stating the name of the chemical involved and the first aid treatment given.

(*ii*) *Mouth*. If the chemical has not been swallowed, the mouth should be rinsed out many times with water but none of the rinsings swallowed. In the event of swallowing a poisonous chemical it should be diluted by the drinking of large amounts of water or milk and the patient immediately removed to hospital. Again information as to the treatment so far administered and the name and estimated amount of chemical involved should accompany the patient to hospital.

The induction of vomiting either by the use of emetics or by tickling the back of the throat should be avoided in most cases.

Whether medical treatment is necessary or not, in the case of splashes on the *skin* or of *inhalation,* can only be judged by the teacher. Inhalation provides a very rapid means of entry of certain chemicals directly to the blood stream whilst for others absorption via the skin is rapid. Area of skin affected, size of dose and variation of an individual's response can make difficult the use of hard and fast rules. Those chemicals which are thought by modern toxicologists to offer a high risk when splashed on the skin or when inhaled will have the advice to seek medical attention, but it does not mean that medical advice and attention should not be sought in other cases.

Ideally all teachers should take advantage of the training and the refresher courses offered by the St. Andrew's Ambulance Association, the St. John Ambulance Association and the Red Cross. The object is to be able to help an injured person before the help of a doctor or nurse can be obtained. A knowledgeable person can by prompt action often save life and prevent an injury from being made worse.

(**p**)   Each page has a heading *Local Conditions* and the space below is left for the teacher to write in information such as the place of storage in his school, any particular recommended means of disposal of that chemical via drains which will satisfy the local water authority, etc.

(**q**)   Finally, the absence of a chemical from the manual does not imply that it is harmless.

[1]Sabre Safety Ltd., Ash Rd., Aldershot, Hampshire GU 4DD.
[2]Draeger Safety Ltd., Draeger House, Sunnyside Rd., Chesham, Bucks HP5 2AR.

Alternative Chemical Names

All chemicals in this manual are listed alphabetically under the ASE recommended names. The following list shows alternative and 'traditional' names in the left-hand column with the corresponding names used in this manual in the right-hand column. To find the information on a particular salt, it may be necessary to look up the page on the anion, e.g. for ammonium fluoride see fluorides.

Alternative or Traditional Name	Recommended Name
acetaldehyde	ethanal
acetic acid	ethanoic acid
acetic anhydride	ethanoic anhydride
acetone	propanone
acetyl chloride	ethanoyl chloride
acetylene	ethyne (see calcium dicarbide and gas cylinders)
allyl bromide	3-bromoprop-1-ene
ammonium sulphocyanate	ammonium thiocyanate
amyl acetate	pentyl ethanoate
iso-amyl alcohol	3-methylbutan-1-ol
n- and sec-amyl alcohols	pentan-1-ol and pentan-2-ol
aniline	phenylamine
aniline hydrochloride	phenylammonium chloride
benzaldehyde	benzenecarbaldehyde
p-benzoquinone	cyclohexa-2,5-diene-1,4-dione
benzoyl chloride	benzenecarbonyl chloride
benzoyl peroxide	di(benzenecarbonyl) peroxide
benzyl alcohol	phenylmethanol
benzyl chloride	(chloromethyl)benzene
bleach	sodium chlorate (I) solution
n-butanol	butan-1-ol
n-butyl acetate	butyl ethanoate
iso-butyl alcohol	2-methylpropan-1-ol
sec-butyl alcohol	butan-2-ol
n-butyl bromide	1-bromobutane
sec-butyl bromide	2-bromobutane
n-butyl chloride	1-chlorobutane
sec-butyl chloride	2-chlorobutane
butyl iodide	1-iodobutane
n-butyric acid	butanoic acid
calcium chlorate (I)	bleaching powder
caradate 30	bis(4-isocyanatophenyl)methane
carbolic acid	phenol
carbon tetrachloride	tetrachloromethane
catechol	benzene-1,2-diol

Alternative or Traditional Name	Recommended Name
caustic potash	potassium hydroxide
caustic soda	sodium hydroxide
charcoal	carbon
chloral hydrate	2,2,2-trichloroethanediol
chloroacetic acids	chloroethanoic acids
chloroform	trichloromethane
chromium trioxide	chromium (VI) oxide
cresols	methylphenols
diaminohexane	hexane-1,6-diamine
diethyl ether	ethoxyethane
diethyl ketone	pentan-3-one
ether	ethoxyethane
ethyl acetate	ethyl ethanoate
ethyl alcohol	ethanol
ethyl bromide	bromoethane
ethyl chloride	chloroethane
ethylene dichloride	1,2-dichloroethane
ethyl iodide	iodoethane
formaldehyde	methanal
formalin	methanal
formic acid	methanoic acid
fuming sulphuric acid	oleum
hexahydrobenzene	cyclohexane
hexahydrophenol	cyclohexanol
hexamethylenediamine	hexane-1,6-diamine
hydroquinone	benzene-1,4-diol
hypochlorites	chlorates (I)
1,2,3-indanetrione hydrate	ninhydrin
iodine trichloride	diiodine hexachloride
iodoform	triiodomethane
lampblack	carbon
lauroyl peroxide	di(dodecanoyl) peroxide
metaldehyde	ethanal tetramer
methyl acetate	methyl ethanoate
methyl alcohol	methanol
methyl bromide	bromomethane
methyl chloroform	1,1,1-trichloroethane
methylene chloride	dichloromethane
methyl ethyl ketone	butanone
methyl methacrylate	methyl 2-methylpropenoate
1- or α-naphthol and 2- or β-naphthol	naphthalen-1-ol and naphthalen-2-ol
nitrates (III)	nitrites
orthophosphoric acid	phosphoric (V) acid
oxalic acid and oxalates	ethanedioic acid and ethanedioates
ozone	trioxygen
paraformaldehyde	poly(methanal)

Alternative or Traditional Name	Recommended Name
paraldehyde	ethanal trimer
perchlorates	chlorates (VII)
perchloric acid	chloric (VII) acid
phenyl bromide	bromobenzene
phenyl chloride	chlorobenzene
p-phenylenediamine	benzene-1,4-diamine
phosphorus (III) chloride	phosphorus trichloride
phosphorus (V) chloride	phosphorus pentachloride
phosphorus pentoxide	phosphorus (V) oxide
phthalic anhydride	benzene-1,2-dicarboxylic anhydride
picric acid	2,4,6-trinitrophenol
potassium ferricyanide	potassium hexacyanoferrate (III)
potassium ferrocyanide	potassium hexacyanoferrate (II)
potassium permanganate	potassium manganate (VII)
potassium persulphate	potassium peroxodisulphate (VI)
powdered graphite	carbon
propionaldehyde	propanal
propionic acid	propanoic acid
n-propyl acetate	propyl ethanoate
n- and iso-propyl alcohol	propan-1-ol and propan-2-ol
n- and iso-propyl chloride	l-chloropropane and 2-chloropropane
propyl iodide	1-iodopropane
pyrogallol	benzene-1,2,3-triol
quicklime	calcium oxide
quinol	benzene-1,4-diol
quinone	cyclohexa-2,5-diene-1,4-dione
resorcinol	benzene-1,3-diol
salicyclic acid	2-hydroxybenzoic acid
sebacoyl chloride	decanedioyl dichloride
slaked lime	calcium hydroxide
sodium hypochlorite solution	sodium chlorate (I) solution
sodium nitrate (III)	sodium nitrite
stannic chloride	tin (IV) chloride, anhydrous
styrene	phenylethene
sulphur monochloride	disulphur dichloride
sulphuryl chloride	sulphur dichloride dioxide
1,2,3,4-tetrahydrobenzene	cyclohexene
thiocarbamide	thiourea
thionyl chloride	sulphur dichloride oxide
toluene	methylbenzene
trichloroethylene	trichloroethene
vinyl acetate	ethenyl ethanoate
xylenes	dimethylbenzenes
xylols	dimethylbenzenes

e.g. Ninhydrin, paints, insecticides.

Hazards In addition to the low BP liquid producing the pressure these are varied in their contents and may be poisonous and flammable. They should be stored away from any sources of heat and probably considered as compressed gases. Pupil use should be under the strictest control; any spraying should be in a well-ventilated area or fume cupboard. Take care if spraying outside on a windy day. There should be no flames or source of possible ignition anywhere near the spraying operation.

Disposal With normal rubbish in bin. Do not incinerate. Can-puncturing appliances are commercially available.

Local Conditions

Fine metal dust.

Hazards Dust is harmful if inhaled. (Possibility of dust explosion when exposed to flame or by chemical action.)

Incompatibility With strong oxidising agents there is a possibility of explosions, e.g. a mixture with oxidising agents including chlorates (V), metal oxides, sulphur, halogens, ammonium nitrate, carbon dioxide, alcohols and silver nitrate.

Handling Powder must be handled carefully to avoid dust rising.

Storage General store with reducing agents.

Disposal Advisable to moisten with water before placing in waste bin.

Spillage Moisten with water, shovel into plastic bag and place in waste bin.

First Aid

Eyes Irrigate with water. Treat as for non-toxic foreign body in the eye. Seek medical advice.

Lungs Remove patient to fresh air.

Mouth Wash out mouth with water.

Skin ——

Local Conditions

$AlCl_3$ Solid, very hygroscopic.

Hazards   Poisonous if taken by mouth, the immediate local reaction causing severe burns. Inhalation of the dust produces irritation or burns of the mucous membranes. The material will cause painful eye burns. When moisture is present on the skin, heat is produced on contact, resulting in thermal and acid burns.

Incompatibility   Water reacts violently with it to form hydrogen chloride.

Handling   Wear gloves and eye protection, use fume cupboard.

Storage   With acids in general store.

Disposal   In fume cupboard, add a little at a time to a large quantity of water. Wash to waste with running water.

Spillage   Wear gloves and face shield. Shovel into dry bucket, transport to safe open site and cautiously add, in small portions, to a large volume of water; after reaction complete run to waste with a large volume of water.

First Aid

Eyes   Irrigate with water. Seek medical attention.

Lungs   Remove patient from exposure; rest and keep warm. Seek medical advice.

Mouth   If swallowed, wash out mouth thoroughly with water and give plenty of water to drink followed by milk of magnesia. Seek medical advice.

Skin   Drench with large quantity of water.

Local Conditions

A Ammonia

NH_3 Pungent smell, lachrymatory.

Hazards   Poisonous if inhaled or swallowed. Gas and solutions irritant to the eyes. Solution burns the skin.
TLV 25 ppm (18 mg m^{-3}).

Incompatibility   Halogens, hydrogen fluoride, mercury, bleaching powder, silver salts.

Handling   Wear gloves and eye protection and handle in fume cupboard. On storage concentrated solution may produce some increased pressure and stopper should therefore be removed slowly with the neck of the bottle facing away from operator.

Storage   In a *cool* cupboard in general store. Only dilute solutions in laboratory. Small quantities of concentrated solutions to be kept only in advanced laboratory.

Disposal   Neutralise and wash to waste with running water.

Spillage   Evacuation of the room is necessary even when the spillage is quite small. Wear gloves, eye protection and respirator. Wash to waste with running water.

First Aid

Eyes   Irrigate with water. Seek medical attention.

Lungs   Remove patient from exposure, rest and keep warm. If exposure large, seek medical advice.

Mouth   Wash with water. If swallowed give plenty of water. Seek medical advice.

Skin   Wash with soap and water. Wash contaminated clothing.

Local Conditions

$(NH_4)_2Cr_2O_7$ Orange solid.

Hazards   Poisonous if taken by mouth. The dust irritates the eyes, skin and respiratory tract. (Fire hazard if heated to decomposition.) A recognised carcinogen (Sax).
Chronic effect: dust produces asthmatic symptoms.
TLV 0.05 mg m^{-3} as Cr.

Incompatibility   Reducing agents, including powdered metals. Explosive mixture with magnesium powder.

Handling   Use gloves and eye protection.

Storage   Keep with oxidising agents. Small quantities in laboratory.

Disposal   Wash to waste with running water.

Spillage   Moisten with water, shovel into bucket and wash to waste with running water. Wash area of spillage well.

First Aid

Eyes   Irrigate with water. Seek medical attention.

Lungs   Remove patient from exposure, rest and keep warm.

Mouth   Wash thoroughly with water. If swallowed, drink large quantities of water. Seek medical advice.

Skin   Wash thoroughly with water. Wash clothing well.

Local Conditions

NH_4NO_3 White crystalline. solid.

Hazards   Dust is irritant to eyes and lungs. Will explode when heated above 250°C and more readily when contaminated. Small quantities only should be heated.

Incompatibility   Dangerous mixtures with acids, chlorates (I, III and V), combustible materials, flammable liquids, nitrites (nitrates (III)), sulphur, carbon, organic compounds and metals (especially if finely divided).

Handling   Wear eye protection. Not to be ground in a mortar or heated strongly.

Storage   With oxidising agents.

Disposal   Wash to waste with water. Do not add to waste bin.

Spillage   As for disposal and wash area of spillage well.

First Aid

Eyes   Irrigate eyes with water. Seek medical advice.

Lungs   Remove patient from area of contamination.

Mouth   Wash with water. Seek medical advice.

Skin   Wash with water.

Local Conditions

$(NH_4)_2S$

Hazards Poisonous and corrosive liquid and vapour. Vapour if inhaled in high concentration may cause unconsciousness. Low concentrations cause headache, giddiness and loss of energy some time after exposure.

Incompatibility With acids and acid fumes hydrogen sulphide is liberated.

Handling Wear gloves and eye protection. Use only in fume cupboard.

Storage General store but not in open laboratory.

Disposal Add to large quantity of water and wash to waste with water.

Spillage Wear gloves and eye protection. Mop up with plenty of water and wash to waste with a large volume of water.

First Aid

Eyes Irrigate with water. Seek medical attention.

Lungs Remove patient from exposure, rest and keep warm.

Mouth If swallowed wash out mouth thoroughly with water. Seek medical attention.

Skin Drench with water. Wash clothing well before re-use.

Local Conditions

Ammonium thiocyanate
Ammonium sulphocyanate

NH_4SCN Yellow crystalline solid.

Hazards Low toxicity but prolonged exposure is harmful. If swallowed is harmful.

Incompatibility Dangerous when heated to decomposition (170°C) or in contact with acids since fumes containing cyanides may be produced.

Handling No special precautions required, but see above.

Storage General store.

Disposal Wash to waste with running water.

Spillage As for disposal.

First Aid

Eyes Irrigate with water. Seek medical advice.

Lungs ——

Mouth Wash with water. Seek medical advice.

Skin Wash with water.

Local Conditions

Hazards Poisonous if inhaled, swallowed or absorbed through the skin. *TLV* 0.5 mg m^{-3}.

Incompatibility A very poisonous gas stibine is formed by the action of acidic reducing agents on antimony and its compounds. Antimony sulphide forms a dangerous mixture with chlorates (V) and chlorates (VII) (perchlorates). Chronic effects by ingestion and inhalation.

Handling If vapour is given off, as is the case with antimony (III) chloride, substances must be handled in a fume cupboard using gloves and eye protection.

Storage Poisons cupboard.

Disposal If quantities very small, wash to waste with water. In the case of antimony (III) chloride do this in a fume cupboard.

Spillage Shovel into a bucket. Consult Local Authority about disposal in an open area. Antimony (III) chloride, since it fumes in moist air, must be dealt with wearing gloves and eye protection. Mop up area of spillage thoroughly with water.

First Aid

Eyes Irrigate with water. Seek medical attention.

Lungs Remove patient from exposure, rest and keep warm. Seek medical advice.

Mouth Wash out mouth thoroughly with water. Seek medical advice.

Skin Wash thoroughly with water.

Local Conditions

A Arsenic and compounds

Schedule 1 poisons.

Hazards Extremely poisonous if inhaled, swallowed or absorbed by skin. Chronic effects by inhalation, by ingestion of dust and by skin contact. A recognised carcinogen (Sax).
TLV as arsenic 0.5 mg m^{-3}.

Incompatibility Arsenic when heated or on contact with acid produces very toxic fumes. Arsenic compounds can react with reducing agents, especially with 'nascent hydrogen' to form arsine gas which is very poisonous.

Handling Avoid raising dust, wear gloves.

Storage Poisons cupboard.

Disposal Collect in suitable container. Consult Local Authority.

Spillage Wearing gloves, respirator and face shield mop up solutions of arsenic compounds into plastic bucket and consult Local Authority. Slightly moisten any solid, carefully sweep up, seal in bag and consult Local Authority.

First Aid

Eyes Irrigate with water. Seek medical advice.

Lungs Remove patient from exposure to dust and fumes; rest and keep warm. Seek medical attention.

Mouth Wash mouth with water. Seek medical attention.

Skin Wash thoroughly with soap and water. Seek medical advice.

Local Conditions

Hazards Dust is irritant to eyes, lungs and skin. Blue asbestos (crocidolite) is carcinogenic, and other forms (chrysotile, amosite) should be avoided where possible. Prolonged inhalation causes tuberculosis and cancer of the lung. Effect can be detected years after a short exposure. Soft forms including 'paper' tape, rope and asbestos wool should not be available in school laboratories. Hard cement forms should be replaced by alternatives where practicable.

Handling Wear gloves and face mask. Handle carefully to avoid dust formation. If possible use in moist condition. Less danger with asbestos/cement board. Soft asbestos or wool most dangerous since dust easily formed.

Disposal In a sealed, stout, well-labelled polythene bag after wetting, and disposed of by contractors in accordance with Deposit of Poisonous Waste Act.

Spillage Wear dust respirator. If small quantity, and if little dust formed, wet and sweep carefully into plastic bag and seal it. If large quantity, or much dust created, or if type of asbestos unknown, the room should be evacuated and specialist cleaners contacted.

First Aid

Contaminated clothing may need to be removed, placed in polythene bags and sealed. Persons contaminated by dust should immediately take a shower or bath. Eyes and mouth should be well irrigated.

Local Conditions

Solid stored under suitable oil.

Hazards Poisonous. Reacts with water to produce highly flammable hydrogen and a poisonous, corrosive solution of barium hydroxide.

Incompatibility Water, air, oxidising agents.

Handling Wear eye protection. Store in air-tight bottle. Barium metal should be kept covered with liquid paraffin or other liquid in which it has been supplied. Great care required that proper liquid is used for topping up. Use tongs or forceps, do not use hands when lifting the metal. Handle on dry filter paper and away from water. Expose to air for minimum time.

Storage Poisons cupboard.

Disposal Small amount up to pea size can be added to water and when reaction is complete, can be washed to waste with water.

Spillage Wear face shield. Cover with dry anhydrous sodium carbonate, shovel into a dry bucket and transfer to an open area. Add to propan-2-ol, leave till all barium has dissolved. Finally run to waste with plenty of water.

First Aid

Eyes Flood with plenty of water. Seek medical attention.

Lungs ——

Mouth Wash mouth with water. Seek medical attention.

Skin Wash with soap and water. Burning possible due to absorption of water from skin. Seek medical attention.

Local Conditions

Schedule 1 poisons.

Hazards Poisonous if swallowed or absorbed through the skin.
TLV 0.5 mg m^{-3}.

Handling Use gloves. Never use mouth suction for pipetting solutions.

Storage Poisons cupboard.

Disposal Consult Local Authority. Washing to waste with lots of water may be permitted. Acid soluble salts, e.g. carbonate, can be dissolved in dilute hydrochloric acid and similarly treated. (Sulphate may be placed in sealed bag with normal waste.)

Spillage As in disposal.

First Aid

Eyes Irrigate with water. Seek medical advice.

Lungs Seek medical advice.

Mouth Wash mouth with water. If swallowed, rest and keep warm. Seek medical attention.

Skin Wash thoroughly with water. Seek medical advice.

Local Conditions

B Barium peroxide

BaO_2 Schedule 1 poison like other soluble barium compounds. Yellow solid.

Hazards Powder is harmful to eyes, lungs, skin. It is fairly corrosive, causing burns. Can form explosive mixtures.

Incompatibility Dangerous if mixed with reducing agents, e.g. ethanoic acid (acetic acid), metal powders or organic substances.

Handling Wear eye protection. Avoid contact with skin by wearing gloves.

Storage Poisons cupboard. Keep dry and stopper tightly.

Disposal Not in waste bin. Small quantities, add a little at a time to running water and wash to waste.

Spillage Wear face shield and gloves. Shovel into a suitable container, e.g. a metal bucket, and if quantity is large consult Local Authority about disposal. Mop the area of spillage with water.

First Aid

Eyes Irrigate with water. Seek medical attention.

Lungs Remove patient from exposure. Rest and keep warm. Seek medical attention.

Mouth Wash with water. Seek medical attention.

Skin Wash thoroughly with water. Seek medical advice.

Local Conditions

Contains sodium carbonate Na_2CO_3, sodium 2-hydroxypropane-1,2,3-tricarboxylate (sodium citrate) $Na_3C_6H_5O_7$, and copper (II) sulphate $CuSO_4$.
Solutions used for quantitative work contain in addition to the above the following: potassium thiocyanate KCNS and a trace of potassium hexacyanoferrate (II) (ferrocyanide) $K_4Fe(CN)_6$.

Hazards Less corrosive than Fehling's Solution. Slightly harmful if swalled owing to presence of copper salts, approximately molar sodium carbonate and the thiocyanate.

Incompatibility Zinc, aluminium.

Handling Wear eye protection.

Storage General store with alkalis.

Disposal Wear gloves and eye protection, dilute with water and wash to waste.

Spillage Wear gloves and eye protection. Mop up and wash to waste.

First Aid

Eyes Irrigate with water. Seek medical attention.

Lungs ——

Mouth Wash thoroughly with water. Give water to drink and if swallowed seek medical attention.

Skin Wash well with water.

Local Conditions

C_6H_6 Liquid, BP 80°C. Benzene should not normally be available in any educational laboratory.

Hazards Highly flammable. Extremely poisonous if swallowed. Harmful if vapour is inhaled. Harmful if vapour or liquid is absorbed through the skin. A recognised carcinogen (Sax). Exposure to low concentrations over prolonged period can cause aplastic leukaemia. Tolerance of individuals cannot be assessed.
(C) *TLV* (skin) 10 ppm (30 mg m^{-3}).
Flash point −11°C.
Autoignition temperature 561°C.

Incompatibility Dangerous if mixed with chlorine, oxidising agents.

Handling Wear gloves and eye protection. Use in fume cupboard.

Storage Flammables store.

Disposal Immiscible with water. Avoid unnecessary discharge to drains. Unavoidable quantities such as washings from glassware etc., should be emulsified and washed to waste.

Spillage Wear respirator, eye protection and gloves. Immediately turn off all sources of ignition in the room. Apply detergent in water, brushing to give an emulsion. Mop and wash the emulsion to waste with plenty of water or spread out in open ground. Wash area of spillage with water and more detergent.

First Aid

Eyes Irrigate with water. Seek medical advice.

Lungs Remove patient from exposure, rest and keep warm. Seek medical advice.

Mouth Wash out mouth thoroughly with water. Seek medical attention.

Skin Wash with soap and water. Remove and air contaminated clothing. Seek medical advice.

Local Conditions

C_6H_5CHO Liquid, BP 179°C. Smells of almonds.

Hazards   Harmful to eyes, lungs, skin. Poisonous by swallowing and skin absorption. Contact may cause dermatitis. Flammable.
Flash point 64°C.
Autoignition temperature 192°C.

Incompatibility   Potassium manganate (VII) (potassium permanganate), sodium peroxide and other oxidising agents.

Handling   In well-ventilated area away from flames and hot plates, etc. Wear gloves and eye protection.

Storage   Flammables store. Up to 250 cm³ in laboratory.

Disposal   Immiscible with water. Unavoidable discharges, e.g. small quantities such as washings from glassware, etc. should be emulsified and washed to waste.

Spillage   Evacuate room and turn off all sources of ignition. Wear face shield and gloves, brush into emulsion with water and detergent. Wash to waste. If detergent not available or spillage is large, absorb in sand, shovel into a bucket, remove to open area for evaporation.

First Aid

Eyes   Irrigate with water. Seek medical advice.

Lungs   Remove patient from area. Keep warm.

Mouth   Wash with water. Seek medical advice.

Skin   Wash with soap and water. Seek medical advice.

Local Conditions

B Benzenecarbonyl chloride
Benzoyl chloride

C_6H_5COCl Liquid, BP 197°C. Lachrymatory.

Hazards Fuming, pungent liquid. Poisonous if swallowed. There is immediate irritation and damage. The vapour irritates the respiratory system and the liquid burns the eyes severely. The liquid is very irritating to the skin and can cause burns.
Flash point 102°C.

Incompatibility Water (reacts to give benzoic acid and hydrochloric acid), alkalis.

Handling Wear gloves and eye protection. Use fume cupboard.

Storage Acids cupboard.

Disposal Add carefully to water, neutralise with alkali solution and run to waste with plenty of water.

Spillage Wear gloves, eye protection and respirator. Apply detergent and work to an emulsion with brush and water – run this to waste diluting greatly. Alternatively, absorb on sand, and transport to safe open area for evaporation. Area of spillage should be washed thoroughly with water and detergent.

First Aid

Eyes Irrigate with water. Seek medical attention.

Lungs If vapour inhaled remove patient from exposure, rest and keep warm. Seek medical advice.

Mouth If swallowed, wash out mouth thoroughly with water and give water to drink followed by milk of magnesia. Seek medical attention.

Skin Drench with water, remove and wash contaminated clothing before re-use. Seek medical advice.

Local Conditions

$C_6H_4(NH_2)_2$ Mauve crystals, MP 140°C.

Hazards Poisonous if swallowed or absorbed by skin. Very irritant to eyes and skin. Harmful if vapour or dust inhaled. Emits toxic fumes when heated. Flammable.
TLV (skin) 0.1 mg m⁻³.
Flash point 156°C.

Incompatibility Oxidising agents.

Handling Wear gloves, and eye protection. Avoid raising dust.

Storage Poisons cupboard.

Disposal Add to excess dilute hydrochloric acid, leave for 24 hours and then wash to waste with running water.

Spillage Wear gloves and face shield, shovel into plastic bucket and add excess dilute hydrochloric acid. Leave for 24 hours, then wash to waste with running water. Wash spillage area with water and detergent.

First Aid

Eyes Irrigate with water. Seek medical attention.

Lungs Remove patient from area, rest and keep warm.

Mouth Wash with water. Seek medical attention.

Skin Wash with soap and water. Remove and wash contaminated clothing.

Local Conditions

Benzene-1,2-dicarboxylic anhydride
Phthalic anhydride

$C_6H_4(CO)_2O$ White crystals, MP 131°C.

Hazards   Irritant to skin and mucous membranes and to alimentary tract if ingested.
TLV 1 ppm (6 mg m^{-3}).

Incompatibility   Oxidising agents including copper (II) oxide, nitric acid and sodium nitrite.

Handling   Wear eye protection and gloves. Avoid raising dust.

Storage   General store.

Disposal   Emulsify and wash small amounts to waste.

Spillage   Wearing gloves and eye protection, slightly dampen and carefully brush up treating as in disposal.

First Aid

Eyes   Irrigate thoroughly with water. Seek medical advice.

Lungs   Remove from exposure.

Mouth   Wash well with water. Seek medical advice.

Skin   Wash well with soap and water.

Local Conditions

$C_6H_4(OH)_2$ White solid, soluble in water, MP 105°C.

Hazards	Very poisonous by ingestion, inhalation and by skin absorption. Burns skin and eyes. Prolonged exposure to low concentrations very harmful. *TLV* 5 ppm (20 mg m^{-3}).
Incompatibility	Oxidising agents especially nitric acid. Emits highly toxic fumes when heated.
Handling	Wear gloves and eye protection.
Storage	Poisons cupboard.
Disposal	Wash small amounts to waste with water.
Spillage	Wearing gloves and face shield, mop up with plenty of water and wash to waste.

First Aid

Eyes	Irrigate well with water. Seek medical attention.
Lungs	Remove from exposure. Rest and keep warm. Seek medical attention.
Mouth	Wash thoroughly with water. If swallowed give plenty of water to drink. Seek medical attention.
Skin	Wash with soap and water. Remove contaminated clothing. Seek medical attention.

Local Conditions

Benzene-1,3-diol
Resorcinol

$C_6H_4(OH)_2$ Colourless to pink solid, MP 110°C.

Hazards   Solutions of the solid readily absorbed by the skin. Solid harmful to the eyes; poisonous if swallowed or by skin absorption if exposed to low concentrations over a long period.
TLV 10 ppm (45 mg m^{-3}).
Flash point 127°C.
Autoignition temperature 342°C.

Incompatibility   Oxidising agents.

Handling   Wear gloves and eye protection.

Storage   General store.

Disposal   Wash to waste with water.

Spillage   Wear gloves and face shield. Mop up with plenty of water and wash to waste with water.

First Aid

Eyes   Irrigate with water. Seek medical attention.

Lungs   Remove patient from exposure, rest and keep warm. Seek medical advice.

Mouth   Wash thoroughly with water. If swallowed give plenty of water to drink. Seek medical advice.

Skin   Wash with soap and water. Remove and wash contaminated clothing. Seek medical advice.

Local Conditions

$C_6H_4(OH)_2$ Solid, MP 170°C.

Hazards Harmful by skin absorption. Dangerous poison. Vapour very harmful to eyes and lungs.
TLV 2 mg m^{-3}.
Flash point 165°C.
Autoignition temperature 516°C.

Incompatibility Oxidising agents, nitric acid.

Handling Wear gloves and eye protection. Handle in fume cupboard.

Storage Poisons cupboard. Never in laboratory.

Disposal Small quantities can be dissolved in a large excess of water and washed to waste with water.

Spillage Wear respirator and gloves. Collect in plastic bag, seal and place in normal refuse. Mop up spillage area with water and wash to waste with running water.

First Aid

Eyes Irrigate with water. Seek medical attention.

Lungs Remove patient from exposure, rest and keep warm. Seek medical attention.

Mouth Wash thoroughly with water. If swallowed give plenty of water to drink. Seek medical attention.

Skin Drench skin with water and wash thoroughly with soap and water. Blisters or burns must receive medical attention. Wash contaminated clothing.

Local Conditions

B Benzene-1,2,3-triol
Pyrogallol

$C_6H_3(OH)_3$ Solid, MP 133°C.

Hazards	Readily absorbed through the skin. Harmful to eyes and lungs.
Incompatability	Oxidising agents.
Handling	Wear gloves and eye protection.
Storage	General store. Never in laboratory.
Disposal	Small quantities can be emulsified with water containing a detergent and washed to waste with water.
Spillage	Small quantities as in disposal. Mop area of spillage with water and detergent and wash to waste with water.

First Aid

Eyes	Irrigate with water. Seek medical attention.
Lungs	Remove patient from exposure, rest and keep warm. Seek medical attention.
Mouth	Wash with plenty of water. If swallowed give plenty of water to drink. Seek medical attention.
Skin	Drench skin with water. Wash contaminated clothing. Seek medical attention.

Local Conditions

These substances should not normally be available in educational establishments. Fluorescent light bulbs contain beryllium and care should be taken if a tube is broken.

Hazards Extremely harmful dust – poisonous if swallowed. Irritates the respiratory system and eyes. Harmful by skin absorption. Suspected carcinogen of lungs and bones (Sax). Cuts and scratches containing particles of beryllium may be slow to heal. Chronic effects by inhalation of dust. Beryllium metal dust or powder will burn if exposed to flame.
TLV 0.002 mg m^{-3}.

Handling Use respirator, eye protection and gloves; avoid raising dust of metals and compounds.

Storage Poisons cupboard.

Disposal Consult Local Authority.

Spillage Consult Local Authority.

First Aid

Eyes Irrigate with water. Seek medical attention.

Lungs Remove patient from exposure, rest and keep warm. Seek medical advice.

Mouth If swallowed, wash out mouth thoroughly with water. Seek medical advice.

Skin Drench with water and remove any penetrating particles. Seek medical attention.

Local Conditions

B   Bis(4-isocyanatophenyl)methane
Caradate 30

$OCNC_6H_4CH_2C_6H_4NCO$ Solid or yellow viscous liquid, used in preparation of polyurethane foam polymer. MP 37°C, BP 194–199°C.

Hazards   Dangerous solid like cyanides. Poisonous if inhaled, swallowed or absorbed through the skin.
TLV 0.02 ppm (0.2 mg m^{-3}).

Incompatibility   Reacts with water.

Handling   Wear gloves and eye protection. Use in a fume cupboard. Should not be heated above 40°C. Any spillage on skin should be washed off immediately.

Storage   In airtight container at temperature below 20°C in outside store if possible or poisons cupboard.

Disposal   In a fume cupboard. Add a little at a time very carefully to large volume of water containing detergent. When reaction complete, wash to waste with running water.

Spillage   Wear gloves, face shield and respirator. Shovel into dry bucket. Consult Local Authority. Mop area of spillage with water and detergent and wash to waste with running water.

First Aid

Eyes   Irrigate with water. Seek medical attention.

Lungs   Remove patient from area of contamination and keep warm. Seek medical advice.

Mouth   Wash out with plenty of water. Seek medical advice.

Skin   Wipe off immediately and wash with detergent and water. Contaminated clothing must be removed and washed. Seek medical advice.

Local Conditions

Hazards	Moderate fire hazard if bismuth is exposed to flame. The element and compound are injurious to eyes or if inhaled or swallowed.
Incompatibility	Oxidising agents, acids and acid fumes.
Handling	Powders should be handled carefully to avoid dust rising. Wear eye protection.
Storage	General store.
Disposal	No special precautions other than eye protection, unless for powdered metal which should be moistened and put in a sealed plastic bag before placing in waste bin.
Spillage	As for disposal.

First Aid

Eyes	Irrigate with water. Seek medical advice.
Lungs	Rest and keep patient warm. Seek medical advice.
Mouth	Wash out mouth with water. Seek medical advice.
Skin	Wash with water.

Local Conditions

B Bleaching powder
Calcium chlorate (I)

CaCl (ClO) $4H_2O$ White solid.

Hazards Poisonous corrosive powder; emits chlorine gas which is poisonous. Powder is harmful to eyes, lungs, mouth, skin. Fire danger by chemical action if in contact with combustible materials. Explosion hazard when the powder is heated (oxygen emitted). Explosion when suddenly heated above 100°C.

Incompatibility Sulphur, primary amines, ammonium salts. With dilute acids and water produces toxic fumes. With concentrated acids emits fumes and may explode.

Handling Well away from flames, etc. Use gloves and eye protection.

Storage Dry, airtight bottle, in general store with oxidising agents.

Disposal Wash small quantities to waste with water.

Spillage Wear gloves and eye protection. Shovel into plastic bucket. Add small portions to a large volume of water and wash to waste. Wash area of spillage well.

First Aid

Eyes Irrigate with water. Seek medical advice.

Lungs If dust or chlorine gas inhaled remove patient from contaminated area. Keep warm and seek medical advice.

Mouth Wash with plenty of water. Seek medical advice.

Skin Wash thoroughly with water.

Local Conditions

Hazards Toxic by inhaling, swallowing or absorption through skin, especially if skin is broken.
TLV trihalides 1 ppm;
borates, tetra, sodium salts, anhydrous (disodium heptaoxotetraborate (III) or borax) 1 mg m^{-3};
borates, tetra, sodium salts, decahydrate (disodium heptaoxotetraborate (III) or borax) 5 mg m^{-3}.

Incompatibility Tribromide, trichloride and trifluoride are very toxic liquids, dangerous when heated to decomposition or in contact with water, acids, oxidising agents and phosphorus.

Handling Powders should be handled carefully to avoid dust rising. The tribromide, trichloride and trifluoride should be handled in a fume cupboard using gloves and eye protection.

Storage General store.

Disposal In waste bin except halogen compounds which should be added slowly to a large quantity of water in a fume cupboard. The solution formed can be washed to waste with water.

Spillage Shovel into waste bin except for halogen compounds. Absorb these in dry sand and shovel into a bucket. For the treatment of the halogen compounds wear gloves, eye protection and respirator. Add soda ash and carefully run to waste with water in small portions at a time.

First Aid

Eyes Irrigate with water. Seek medical advice.

Lungs Rest and keep patient warm.

Mouth Wash out mouth with water. Seek medical advice.

Skin Wash with water.

Local Conditions

B Bromine and bromine water

Br_2 Red-brown volatile liquid, BP 59°C.

Hazards   Vapour is extremely irritant to eyes, lungs and skin. Liquid bromine and bromine water are poisonous and corrosive if swallowed. Liquid bromine is extremely corrosive to the skin. Bromine water is much less corrosive and hazardous. *TLV* 0.1 ppm (0.7 mg m^{-3}).

Incompatibility   Dangerous if mixed with ammonia, hydrocarbons, hydrogen, powdered metals, turpentine, ethoxyethane (diethyl ether), methanol and propanone (acetone).

Handling   Wear gloves and eye protection and handle in a fume cupboard. The outside of the bottle should be washed with water after using to remove any bromine that may have run down the side during pouring. A teat pipette can be used for transferring small quantities; this is much safer than pouring from the bottle.

Storage   Bottles and ampoules with acids in general store. If ampoules are used, which is preferable, strict attention should be paid to instructions for opening.

Disposal   Add to strong alkali solution and pour down the drain with plenty of running water. This should be done in a fume cupboard.

Spillage   Evacuate the room. Wear eye protection, respirator and gloves. Add anhydrous sodium carbonate to spillage and mop up with plenty of water. Wash to waste with plenty of running water.

First Aid

Eyes   Irrigate with water. Seek medical attention.

Lungs   Remove patient from exposure. Seek medical advice.

Mouth   Wash out mouth thoroughly with water. Give large quantity of water to drink. Seek medical attention.

Skin   Wash with plenty of water and then with dilute ammonia or a mild reducing agent, e.g. sodium thiosulphate. Seek medical advice.

Local Conditions

C_6H_5Br Liquid, BP 156°C.

Hazards Poisonous if swallowed. Poisonous by skin absorption. The vapour may be narcotic in high concentrations. Irritates the eyes. Flammable.
Flash point 50°C.
Autoignition temperature 566°C.

Incompatibility Oxidising agents.

Handling Wear gloves and eye protection and use fume cupboard if chemical to be exposed for more than a minute or so.

Storage Flammables store. Up to 250 cm³ in laboratory.

Disposal Unavoidable discharges, e.g. small quantities such as washings from glassware, etc. should be emulsified and washed to waste.

Spillage Turn off all sources of ignition and evacuate room. Wearing respirator, eye protection and gloves, brush to an emulsion with water and detergent. Alternatively soak up on sand, collect into a bucket and carry to a safe open space for evaporation. The site of the spillage should be washed well with water.

First Aid

Eyes Irrigate with water. Seek medical advice.

Lungs If vapour inhaled, remove patient from exposure. Rest and keep warm.

Mouth If swallowed wash out mouth thoroughly with water. Seek medical advice.

Skin Drench with water and wash thoroughly with soap and water. Remove and air contaminated clothing before re-use.

Local Conditions

B 1-Bromobutane and 2-bromobutane
n- and sec-Butyl bromide

$CH_3(CH_2)_2CH_2Br$ Liquid, BP 101°C.
$CH_3CHBrCH_2CH_3$ Liquid, BP 91°C.

Hazards   Harmful to eyes, lungs, and if swallowed. Toxicity is unknown. When heated to decomposition emit toxic fumes. Highly flammable.
Flash point 22°C for both.
Autoignition temperature 265°C for 1-bromobutane and about 265°C for 2-bromobutane.

Incompatibility   Oxidising agents.

Handling   In well-ventilated area. Wear eye protection. Do not distil to dryness.

Storage   Flammables store. Up to 250 cm³ in laboratory.

Disposal   Unavoidable discharges, e.g. small quantities such as washings from glassware, etc. should be emulsified and washed to waste.

Spillage   Wear gloves, eye protection and respirator. Brush with water and detergent to form emulsion. Wash to waste with plenty of water.

First Aid

Eyes   Irrigate with water. Seek medical advice.

Lungs   Remove patient from area, rest and keep warm. Seek medical advice.

Mouth   Wash with water. Seek medical advice.

Skin   Wash with soap and water. Seek medical advice.

Local Conditions

C_2H_5Br Liquid, BP 38°C.

Hazards Harmful by inhalation, swallowing and eye or skin contact. When heated to decomposition emits toxic fumes. Less toxic than bromomethane. Chronic effects from exposure to low concentrations.
TLV 200 ppm (890 mg m^{-3}).
Autoignition temperature 517°C.

Incompatibility Oxidising agents including hydrogen peroxide, ammonium nitrate, chromic (VI) acid, nitric acid, oxygen, chloric (VII) acid (perchloric acid) and chlorates (VII) (perchlorates).

Handling In fume cupboard away from sources of ignition. Wear gloves and eye protection. Do not distil to dryness.

Storage Flammables store. Up to 250 cm³ in laboratory.

Disposal Unavoidable discharges, e.g. small quantities such as washings from glassware, etc. should be emulsified and washed to waste.

Spillage Evacuate room and wear respirator, eye protection and gloves. Absorb in sand for removal to open area for evaporation. Wash spillage area with detergent and water.

First Aid

Eyes Irrigate with water. Seek medical attention.

Lungs Remove patient from exposure, rest and keep warm. Seek medical advice.

Mouth Wash out with water. Seek medical advice.

Skin Wash with water and then soap and water. Remove and wash contaminated clothing. Seek medical advice.

Local Conditions

Bromomethane
Methyl bromide

CH_3Br Schedule 1 poison. Gas, BP 4°C.

Hazards   Harmful vapour. Vapour or liquid causes skin irritation. Prolonged exposure to low concentrations harmful to lungs.
TLV (skin) 15 ppm (60 mg m^{-3}).
Autoignition temperature 537°C.

Incompatibility   Oxidising agents.

Handling   Only in fume cupboard, away from sources of ignition. Wear gloves and eye protection. May be supplied in glass ampoule which must be cooled in ice before opening.

Storage   Locked up with flammables. Never in laboratory.

Disposal   Allow to evaporate in fume cupboard.

Spillage   Evacuate room, wear respirator, eye protectors and gloves, and ventilate the area.

First Aid

Eyes   Irrigate with water. Seek medical attention.

Lungs   Remove patient from exposure, rest and keep warm. Seek medical advice.

Mouth   Wash out with water and give plenty of water to drink. Seek medical attention.

Skin   Drench with water and wash thoroughly with soap and water. Blisters or burns *must* receive medical attention. Remove and wash contaminated clothing.

Local Conditions

$CH_2:CHCH_2Br$ Liquid, BP 71°C. Lachrymatory.

Hazards Corrosive liquid; poisonous if swallowed. Harmful by skin absorption. Vapour and liquid irritate the skin, eyes and respiratory system and may cause dizziness and headache. Highly flammable.
Flash point −1°C.
Autoignition temperature 295°C.

Incompatibility Oxidising agents.

Handling Wear gloves and eye protection. Handle in fume cupboard.

Storage Flammables store. Never in laboratory.

Disposal Unavoidable discharges, e.g. small quantities such as washings from glassware, etc. should be emulsified and washed to waste.

Spillage Evacuate the area and turn off all possible sources of ignition. Wearing gloves, eye protection and respirator, brush to an emulsion with detergent and water. Wash this to waste with large amounts of water. Alternatively soak up on sand, lift into a bucket and carry to a safe outdoor site for evaporation. Wash area of spillage with water and detergent.

First Aid

Eyes Irrigate with water. Seek medical attention.

Lungs Remove patient from exposure, rest and keep warm. Seek medical advice.

Mouth If swallowed wash out mouth thoroughly with water. Seek medical advice.

Skin Drench with water and wash with soap and water; remove and wash contaminated clothing before re-use. Seek medical attention.

Local Conditions

B Butanoic acid
n-Butyric acid

$CH_3CH_2CH_2COOH$ Liquid, BP 163.5°C.

Hazards Corrosive action on skin.
Flash point 72°C.
Autoignition temperature 452°C.

Incompatibility Oxidising agents.

Handling Handle carefully using gloves and eye protection.

Storage Flammables store. Up to 250 cm³ in laboratory.

Disposal Run to waste diluting greatly with running water.

Spillage Shut off all possible sources of ignition. Wear gloves. Mop up with plenty of water. Ventilate spillage area well to evaporate remaining liquid and dispel vapour.

First Aid

Eyes Irrigate with water. Seek medical advice.

Lungs Remove patient from exposure, rest and keep warm.

Mouth Wash thoroughly with sodium hydrogencarbonate solution. Give milk of magnesia then plenty of water or milk to drink. Seek medical advice.

Skin Wash thoroughly with water and dilute solution of sodium hydrogencarbonate then thoroughly wash with soap and water.

Local Conditions

$CH_3CH_2CH_2CH_2OH$ Liquid, BP 118°C.
$CH_3CH_2CHOHCH_3$ Liquid, BP 95°C.

Hazards Harmful vapour. Liquid harmful to skin. Liquid can be absorbed by skin causing internal injury. Highly flammable.
(C) *TLV* (skin) 50 ppm (150 mg m^{-3}) for butan-1-ol.
TLV 150 ppm (450 mg m^{-3}) for butan-2-ol.
Flash point 29°C for butan-1-ol and 24°C for butan-2-ol.
Autoignition temperature 365°C for butan-1-ol and 406°C for butan-2-ol.

Incompatibility Oxidising agents.

Handling In fume cupbard preferably; if not available in *well*-ventilated room away from all sources of ignition. Wear gloves and eye protection.

Storage Flammables store. Up to 250 cm^3 in laboratory.

Disposal Unavoidable discharges, e.g. small quantities such as washings from glassware, etc. should be emulsified and washed to waste.

Spillage Turn off all sources of ignition. Wear face shield and gloves. Emulsify and run to waste, or absorb on sand, shovel into buckets. Transfer to open area for evaporation. Spillage area to be thoroughly washed and ventilated.

First Aid

Eyes Irrigate with cold water. Seek medical advice.

Lungs Remove patient from exposure, rest and keep warm.

Mouth Wash thoroughly with cold water. Seek medical advice.

Skin Wash thoroughly with cold water and soap. Remove and wash contaminated clothing.

Local Conditions

Butanone
Methyl ethyl ketone

$CH_3COCH_2CH_3$ Liquid, BP 80°C.

Hazards   Vapour harmful to eyes, lungs. Poisonous if swallowed. Mildly irritant to skin. Highly flammable.
TLV 200 ppm (590 mg m^{-3}).
Flash point −5.6°C.
Autoignition temperature 515°C.

Incompatibility   Oxidising agents, trichloromethane (chloroform).

Handling   In well-ventilated area, well away from flames, etc. Wear eye protection.

Storage   Flammables store.

Disposal   Mix with water and detergent and wash to waste.

Spillage   As for disposal but spraying the spillage with water and detergent, mopping up and washing to waste with large quantity of water.

First Aid

Eyes   Wash with water. Seek medical advice.

Lungs   Remove patient from area and keep warm.

Mouth   Wash out mouth. Seek medical advice.

Skin   Wash with soap and water.

Local Conditions

$CH_3COO(CH_2)_3CH_3$ BP 125°C.

Hazards Harmful vapour. Irritant to eyes and nose. Harmful if swallowed and by skin absorption. Highly flammable.
TLV 150 ppm (710 mg m^{-3}).
Flash point 23°C.
Autoignition temperature 421°C.

Incompatibility Oxidising agents.

Handling In a fume cupboard or in well-ventilated area away from source of ignition.

Storage Flammables store. Up to 250 cm^3 in laboratory.

Disposal Unavoidable discharges, e.g. small quantities such as washings from glassware, etc. should be emulsified and washed to waste.

Spillage Turn off all sources of ignition. Evacuate the room. Wear face shield and gloves. Emulsify and run to waste, or absorb on sand, shovel into bucket for removal to open area for evaporation. Spillage area to be thoroughly ventilated.

First Aid

Eyes Irrigate with water. Seek medical advice.

Lungs Remove patient from exposure, rest and keep warm.

Mouth Wash thoroughly with water. Seek medical advice.

Skin Wash skin with plenty of water. Remove and wash contaminated clothing.

Local Conditions

C Cadmium and compounds

Hazards	Vapours, solids and liquids can be very harmful to eyes, lungs, skin, and by swallowing. Cadmium oxide fumes can cause metal fume fever. Brief exposure to high concentrations of dust or fumes may be fatal. Exposure to low concentrations over a long period results in severe damage to lungs, liver and kidneys. A suspected carcinogen (Sax). *TLV* 0.05 mg m^{-3}.
Incompatibility	Cadmium metal with oxidising agents.
Handling	Solids: handle carefully, with eye protection and gloves. If vapour or dust formed, handle in fume cupboard.
Storage	Poisons cupboard.
Disposal	Very small amounts such as washings may be washed to waste. For large amounts consult Local Authority.
Spillage	Wear dust respirator, eye protection and gloves. Moisten with water, brush up into bucket or bag and consult Local Authority.

First Aid

Eyes	Irrigate with water. Seek medical attention.
Lungs	Remove patient from exposure, rest and keep warm. Seek medical attention.
Mouth	Wash with water. Seek medical attention.
Skin	Wash out thoroughly with water. Seek medical advice.

Local Conditions

Metallic solid.

Hazards Harmful to eyes, lungs and skin. Fumes from burning calcium are composed of calcium oxide (quicklime).

Incompatibility Acids. Reacts with water to form hydrogen. A dangerous mixture with oxidising agents.

Handling Use forceps or spatula, and make sure bench is dry. Wear eye protection.

Storage General store in airtight bottle, with reducing agents.

Disposal Wear eye protection. Add to water in a trough or basin and leave till reaction is complete. Wash to waste in a well-ventilated area to prevent accumulation of hydrogen.

Spillage As for disposal because only small quantities will be involved.

First Aid

Eyes Flood with plenty of water. Seek medical attention.

Lungs ——

Mouth Wash out mouth with water. Seek medical advice.

Skin Wash thoroughly with water.

Local Conditions

CaC_2 Solid.

Hazards Solid is not dangerous. Hazards are due to ethyne (acetylene) and calcium hydroxide formed when in contact with water or acids. Ethyne forms explosive mixture with air.

Incompatibility Water, dilute acids, chlorine, bromine, iodine, any halogen oxoanions, silver nitrate solutions, copper compounds and other heavy metals.

Handling Wear gloves and eye protection. Handle only on dry bench away from water.

Storage Preferably in an airtight can and not a bottle. General store. Not in laboratory.

Disposal In a well-ventilated part of the laboratory or outside, add a little at a time to water. When reaction is complete the residue of calcium hydroxide can be washed down the drain.

Spillage Shovel into a dry bucket and treat the solid as in disposal but outside in an open area.

First Aid

Eyes Irrigate with water. Seek medical attention.

Lungs ——

Mouth Wash with water. Seek medical advice.

Skin Wash thoroughly with water.

Local Conditions

CaH_2 White solid.

Hazards   Harmful to eyes, lungs, mouth and skin.

Incompatibility   Powder can form dust cloud which is explosive if in contact with flame, spark, heat or oxidising agents. Reacts with water to form hydrogen and calcium hydroxide. Heated with water or acids, an exothermic reaction evolving hydrogen occurs and temperature may be high enough to cause ignition.

Handling   Wear gloves and eye protection. Keep away from water.

Storage   Well-stoppered bottle in a dry place. General store with reducing agents.

Disposal   Wear face shield in a well-ventilated area and add a little at a time to large volume of water using spatula in well stretched out hand. Leave for one hour or so for reaction to be complete and wash to waste with water.

Spillage   Wear respirator and face shield, cover with dry anhydrous sodium carbonate, shovel into dry bucket. In an open area add the mixture a little at a time to a large volume of water and stand back after each addition. Allow to stand 24 hours and wash down the drain with running water.

First Aid

Eyes   Irrigate with water. Seek medical advice.

Lungs ——

Mouth   Wash out mouth with water. Large quantities of water to drink. Seek medical advice.

Skin   Wash thoroughly with water.

Local Conditions

CaO and $Ca(OH)_2$ White solids.

Hazards Harmful to eyes and skin, and lungs.
TLV 5 mg m^{-3}.

Incompatability Calcium oxide is dangerous on contact with water, steam, acids or acid fumes since it reacts to produce heat. A lump of calcium oxide may disintegrate violently when water is added to it.

Handling Wear eye protection when adding water to calcium oxide. Preparation of lime water is best done using calcium hydroxide. Wear rubber gloves when handling either solid.

Storage General store with alkalis.

Disposal Calcium oxide by adding it in small portions to a large volume of water, stirring and washing to waste. Calcium hydroxide by washing to waste with water.

Spillage Wear face shield and gloves. Brush up carefully and collect in dry plastic bucket. Carry to open area and treat as in disposal.

First Aid

Eyes Irrigate with water. Seek medical attention.

Lungs Remove patient from area, rest and keep warm. Seek medical advice.

Mouth Wash out mouth with water. If swallowed give large quantity of water to drink. Seek medical advice.

Skin Wash thoroughly with water.

Local Conditions

Hazards	Dust is irritant if inhaled or goes into eyes. Slight explosion hazard in the form of dust when exposed to heat or flame. Charcoal blocks used for oxide reduction can cause fires if stored away after use without ensuring that area used in reduction is properly cooled. *TLV* 3.5 mg m^{-3} (lampblack).
Incompatibility	Oxidising agents.
Handling	Normal chemical laboratory care to prevent accidental spillage and scattering of dust.
Storage	General store with reducing agents.
Disposal	Moisten and put in normal waste bin.
Spillage	Moisten. Sweep carefully into shovel and put in waste bin.

First Aid

Eyes	If particles are in contact with eyes, irrigate with water.
Lungs	Remove patient from area of contamination.
Mouth	Wash out with water.
Skin	Wash with soap and water.

Local Conditions

C Carbon Dioxide (gas and solid)

Hazards	Gas is simple asphyxiant. *TLV* 5000 ppm (9000 mg m^{-3}). Solid (carbon dioxide snow) in contact with skin may cause a burn.
Incompatibility	Potassium, lithium, sodium, metal hydrides, finely divided aluminium, magnesium and tin.
Handling	Gloves and eye protection should be worn when handling the solid. The solid must not be placed in a closed container since danger of explosion due to gas formed from the solid.
Storage	General store. The solid can only be stored for a short time by wrapping in paper or other insulating material, care being taken that any gas formed can escape.
Disposal	The solid by leaving in a well-ventilated space.
Spillage	Shovel into bucket and leave in an open area to evaporate.

First Aid

Eyes	Irrigate with water. Seek medical advice.
Lungs	Remove patient from area of contamination.
Mouth	——
Skin	Solid evaporates very quickly but if skin is burned seek medical advice.

Local Conditions

CS_2 Liquid, BP 46°C. The substance should not normally be available in educational laboratories.

Hazards Vapour is extremely harmful to the eyes and lungs. Little effect on skin due to its rapid evaporation. Liquid is poisonous if swallowed. Affected by chronic exposure are the central nervous system, memory and personality. Highly flammable vapour. Alternative solvents which may be used to prepare rhombic sulphur are warm olive oil, dimethylbenzene, methylbenzene. The crystals are small and a microscope or hand lens is needed.
TLV (skin) 20 ppm (60 mg m⁻³).
Flash point −30°C.
Autoignition temperature 100°C.

Incompatibility Sodium peroxide, oxidising agents.

Handling Small amounts could be used in fume cupboard for teacher demonstrations. Wear gloves and eye protection, use in a well-ventilated area with all sources of ignition in the room turned off. Notice the low autoignition temperature and hence the need to make sure that the vapour does not come in contact with a very hot surface, e.g. electric storage heater, hotplate, hot tripod, gauze, light bulb, etc.

Storage Flammables store.

Disposal Allow small amounts to evaporate in fume cupboard. Emulsify washings from glassware and wash to waste with plenty of water.

Spillage Wear respirator, eye protection and gloves. Again check absence of possible sources of ignition. Apply detergent and water and brush to emulsify the carbon disulphide. Mop up and spread out in an open area for evaporation.

First Aid

Eyes Irrigate with water. Seek medical attention.

Lungs Remove patient from exposure, rest and keep warm. Seek medical advice.

Mouth Wash out well with water. Seek medical attention.

Skin Little harm to skin since evaporation is very rapid. Wash well and seek medical advice.

Local Conditions

ClO^-

Hazards Harmful to skin and eyes. Inhalation of dust or swallowing of compounds very harmful. In contact with acids, even carbon dioxide of the air, chlorine is evolved. Powerful oxidising agents particularly at high temperatures. Hence fire hazard if in contact with combustible materials.

Incompatibility Dilute acids and even water/air yields chlorine gas. With concentrated acids the reaction may be explosive. Sulphur, primary amines, ammonium salts, methanol, benzonitrile and other reducing agents.

Handling Wear gloves and eye protection. Avoid raising dust. Do not warm.

Storage In dry airtight bottle. With oxidising agents.

Disposal Wash small amounts to waste with water.

Spillage Wear gloves and eye protection. Shovel carefully into plastic bucket, add in small portions to a large volume of water and wash to waste. Wash spillage area well.

First Aid

Eyes Irrigate well with water. Seek medical attention.

Lungs If dust or chlorine inhaled remove patient from contaminated area. Keep warm and seek medical advice.

Mouth Wash with plenty of water. Seek medical attention.

Skin Wash thoroughly with water.

Local Conditions

ClO_3^-

Hazards Generally poisonous solids, sodium chlorate (V) is used as weed killer. Harmful to skin and eyes. Very poisonous if swallowed. Dust irritates the respiratory system. Extremely dangerous since if even only slightly contaminated they can explode by moderate shock or when heated to decomposition. Extremely dangerous mixtures with substances listed under *Incompatibility*.

Incompatibility Reducing agents including ammonium salts, antimony sulphide, carbon, combustible materials, finely divided materials, metal powders, sawdust, sugar, sulphur, phosphorus; and with concentrated sulphuric acid. Only pure manganese (IV) oxide should be used with potassium chlorate (V) for oxygen preparation but the method is not recommended. A safer method of preparing oxygen is by catalytic decomposition of 20 volume or 10 volume hydrogen peroxide solution.

Handling Wear eye protection. Keep away from all incompatible materials and sources of heat.

Storage With oxidising agents.

Disposal Run to waste greatly diluted with water.

Spillage Wear face shield. Carefully sweep up with soft brush after moistening with water. Add to plastic bucket containing water. Wash down drain with large volume of water. Site of spillage should be carefully washed with water.

First Aid

Eyes Irrigate with water. Seek medical attention.

Lungs ——

Mouth Rinse thoroughly with water. Seek medical attention.

Skin Wash with water, then with soap and water. Seek medical advice especially if skin is broken.

Local Conditions

ClO_4^-

Hazards	Harmful to skin and eyes. Very poisonous if swallowed. Dust irritates the respiratory system.
Incompatibility	Explosive when mixed with combustible materials. Slight friction can explode such mixtures. Avoid mixing with reducing agents, sugar, sulphur, charcoal, graphite, and sulphuric acid.
Handling	In the open laboratory away from combustible materials and reducing agents. Wear gloves and eye protection.
Storage	General store with oxidising compounds.
Disposal	Dissolve in water and wash to waste with running water.
Spillage	Wear face shield. Wet with water, mop or brush up and put down the drain with plenty of running water. Area should be repeatedly washed with water since contaminated wood, paper and cloth are easily ignited. If spillage large, the contaminated wooden fabric should be replaced.

First Aid

Eyes	Irrigate with water. Seek medical attention.
Lungs	Remove patient from area, rest and keep warm.
Mouth	Wash with water, give water to drink. Seek medical attention.
Skin	Wash with water.

Local Conditions

$HClO_4$ Colourless liquid.

Hazards   Fumes and liquid burn eyes and skin severely. Causes severe irritation if taken into the mouth. It should not be present in establishments which do not possess a specially designed chloric (VII) acid fume cupboard.

Incompatibility   Can cause fire or explosion when in contact with reducing agents, organic substances and finely divided metals. Heating decomposes the acid to give toxic fumes.

Handling   Greatest care required. Handle only in fume cupboard especially designed for chloric (VII) acid. Wear gloves and eye protection. Keep away from incompatible materials, including paper tissues, cloth dusters and wood.

Storage   In the special fume cupboard labelled 'Chloric (VII) Acid Only'.

Disposal   Small quantities can be added to a large volume of water, neutralised with sodium carbonate and washed to waste with running water.

Spillage   Wear gloves and face shield. Spread anhydrous sodium carbonate (soda ash) liberally over the liquid and mop up carefully with plenty of water. Wash to waste with large quantity of running water. Area should be repeatedly washed with water, since contaminated wood, paper and cloth are easily ignited. If spillage is large the wooden fabric should be replaced.

First Aid

Eyes   Irrigate with water. Seek medical attention.

Lungs   Remove patient from area, keep warm. Seek medical advice.

Mouth   Wash out with water and give water to drink, followed by milk of magnesia. Seek medical attention.

Skin   Wash with water and apply paste of glycerol and magnesia.

Local Conditions

Cl_2 Green gas.

Hazards   Very poisonous. Chlorine water if concentrated emits chlorine. Extremely harmful to the eyes, respiratory tract and lungs.
Detectable odour at 3.5 ppm.
Immediate irritation of throat at 15 ppm.
Even short exposure dangerous at 50 ppm.
Brief exposure may be fatal at 1000 ppm.
TLV 1 ppm (3 mg m^{-3}).

Incompatibility   Can react and cause fire or explosion with turpentine, ethoxyethane (diethyl ether), ethyne (acetylene), ammonia gas, coal gas, natural gas, hydrogen and hydrides and powdered metals.

Handling   Always in a fume cupboard. Only dilute chlorine water or bleach should be used in open area. Chlorine cylinders are not recommended in schools.

Storage ——

Disposal   By strong draught in fume cupboard. Chemicals from which chlorine has been prepared should be left in fume cupboard until reaction is complete and all the chlorine has gone. The chemicals should then be washed to waste with plenty of running water.

Spillage   This only applies to spillage of chemicals which react to form chlorine. Evacuate the laboratory, wear respirator and eye protection and add water to dilute and stop reaction. Mop up and wash to waste with water.

First Aid

Eyes   Irrigate with water. Seek medical advice.

Lungs   Remove patient from exposure, rest and keep warm. Seek medical advice.

Mouth   Wash with water. Seek medical advice.

Skin   Wash with plenty of water.

Local Conditions

C_6H_5Cl Liquid, BP 132°C.

Hazards Harmful vapour. Harmful by skin contact. Poisonous by swallowing, inhaling and skin absorption. Highly flammable.
TLV 75 ppm (350 mg m^{-3}).
Flash point 29°C.
Autoignition temperature 638°C.

Incompatibility Oxidising agents and reactive metals.

Handling In fume cupboard, wearing gloves and eye protection.

Storage Flammables store.

Disposal Unavoidable discharges, e.g. small quantities such as washings from glassware, etc. should be emulsified and washed to waste.

Spillage Turn off all possible sources of ignition and evacuate the laboratory. Wearing respirator and gloves, brush to an emulsion with detergent and water. Wash to waste with large amounts of water. Alternatively absorb on sand and transfer to an outdoor site for evaporation. Wash site of spillage well.

First Aid

Eyes Irrigate with water. Seek medical advice.

Lungs Remove patient from exposure, rest and keep warm.

Mouth Wash out mouth thoroughly with water. Seek medical advice.

Skin Drench with water. Wash with soap and water. Remove contaminated clothing and wash *thoroughly* before re-use.

Local Conditions

1-Chlorobutane and 2-chlorobutane
n- and sec-Butyl chloride

$CH_3(CH_2)_2CH_2Cl$ Liquid, BP 78°C.
$CH_3CHClCH_2CH_3$ Liquid, BP 69°C.

Hazards Highly flammable. Dangerous when heated to decomposition emitting carbonyl chloride (phosgene).
Flash point −9°C for 1-chlorobutane and 21°C for 2-chlorobutane.
Autoignition temperature 460°C for both.

Incompatibility Oxidising agents and reactive metals.

Handling In fume cupboard, wearing gloves and eye protection. Do not distil to dryness.

Storage Flammables store.

Disposal Unavoidable discharges, e.g. small quantities such as washings from glassware, etc. should be emulsified and washed to waste.

Spillage Turn off all possible sources of ignition. Instruct others to keep at a safe distance. Wear respirator, eye protection and gloves. Brush to an emulsion with water and detergent. Alternatively, absorb on sand, and take to open area for evaporation. Spillage area should be well washed with detergent and water.

First Aid

Eyes Irrigate with water. Seek medical attention.

Lungs Remove patient from exposure, rest and keep warm.

Mouth Wash out mouth thoroughly with water. Seek medical advice.

Skin Drench with water. Wash with soap and water. Remove contaminated clothing and *thoroughly air* before re-use.

Local Conditions

Chloroethane C
Ethyl chloride

C_2H_5Cl Gas at room temperature, BP 12°C.

Hazards Harmful vapour. Highly flammable.
TLV 1000 ppm (2600 mg m^{-3}).
Flash point −50°C.
Autoignition temperature 519°C.

Incompatibility Oxidising agents and reactive metals.

Handling In fume cupboard, wearing gloves and eye protection.

Storage Flammables store. Never in open laboratory.

Disposal If liquid, allow small quantities to evaporate in fume cupboard.

Spillage Turn off all possible sources of ignition. Instruct others to keep at a safe distance. Wear respirator, eye protection and gloves. Brush to an emulsion with detergent and water. Alternatively absorb on sand, shovel into bucket and take to open area for evaporation. Spillage area should be washed well with detergent and water.

First Aid

Eyes Irrigate with water. Seek medical advice.

Lungs Remove patient from exposure, rest and keep warm.

Mouth Wash out mouth thoroughly with water. Seek medical advice.

Skin Drench with water. Wash with soap and water. Seek medical attention. Remove contaminated clothing and thoroughly air before re-use.

Local Conditions

C Chloroethanoic acids
Chloroacetic acids

$CH_2ClCOOH$ Liquid, BP 189°C.
$CHCl_2COOH$ Liquid, BP 194°C.
CCl_3COOH Solid, MP 57.5°C.

Hazards Corrosive and poisonous liquids. Vapour and liquid harmful to eyes, skin and lungs. Emit fumes of carbonyl chloride (phosgene) and chlorides when heated to decomposition.
TLV not available.

Incompatibility Oxidising agents and reactive metals.

Handling In fume cupboard wearing gloves and eye protection. Do not distil to dryness.

Storage With acids in general store. Never in open laboratory.

Disposal Run to waste diluting greatly with running water.

Spillage Wear gloves and eye protection (rubber boots if spillage is large). Spread soda ash over spillage to neutralise. Mop up cautiously with plenty of water and wash to waste with large quantity of water.

First Aid

Eyes Irrigate with water. Seek medical attention.

Lungs Remove patient from area. Rest and keep warm. Seek medical advice.

Mouth Wash mouth thoroughly with water. Give plenty of water to drink followed by milk of magnesia. Seek medical advice.

Skin Drench skin with water. Obtain medical attention for blisters or burns. Remove contaminated clothing and wash before further use.

Local Conditions

$C_6H_5CH_2Cl$ Liquid, BP 179°C. Lachrymatory.

Hazards Poisonous if swallowed, causing severe internal irritation and damage. The vapour irritates the respiratory system, eyes and skin. The liquid is corrosive. Flammable.
TLV 1 ppm (5 mg m⁻³).

TLV 1 ppm (5 mg m^{-3}).
Flash point 60°C.
Autoignition temperature 585°C.

Incompatibility Oxidising agents. Can decompose violently if heated with metals.

Handling Wear gloves and eye protection; use fume cupboard.

Storage Flammables store. Never in laboratory.

Disposal Unavoidable discharges, e.g. small quantities such as washings from glassware, etc. should be emulsified and washed to waste.

Spillage Wear eye protection and respirator, put out all flames, hot plates, etc. Add water and detergent and brush into an emulsion. Wash to waste with plenty of running water. Alternatively, absorb on sand and transport outside to safe open place for evaporation or burning. Wash area of spillage well with water and detergent.

First Aid

Eyes Irrigate with water. Seek medical attention.

Lungs If vapour inhaled remove patient from exposure, rest and keep warm.

Mouth If swallowed wash out mouth thoroughly with water. Seek medical advice.

Skin Thoroughly soak with water and wash with soap and water. Remove and wash contaminated clothing before re-use.

Local Conditions

C 1-Chloropropane and 2-chloropropane
n- and iso-Propyl chloride

$CH_3CH_2CH_2Cl$ Liquid, BP 45°C.
$CH_3CHClCH_3$ Liquid, BP 36°C.

Hazards   Dangerous when heated to decomposition emitting carbonyl chloride (phosgene). Highly flammable.
Flash point −18°C for 1-isomer and −32°C for 2-isomer.
Autoignition temperature 520°C for 1-isomer and 590°C for 2-isomer.

Incompatibility   Oxidising agents.

Handling   In well-ventilated place, preferably fume cupboard, all flames, etc. off. Wear gloves and eye protection.

Storage   Flammables store.

Disposal   Small quantities such as washings from glassware, etc. should be emulsified and washed to waste.

Spillage   Turn off all flames, hot plates, ovens, etc. Evacuate room. Wearing respirator, eye protection and gloves, emulsify with water and detergent. Mop up and transfer to bucket. If quantity is large spread out in an open area for evaporation. Wash the spillage area throughly.

First Aid

Eyes   Irrigate with water. Seek medical attention.

Lungs   Remove patient from area, rest and keep warm.

Mouth   Wash with water. Seek medical attention.

Skin   Wash with soap and water.

Local Conditions

$ClSO_3H$ A pale yellow colourless liquid, BP 151°C.

Hazards Both vapour and liquid can cause acute toxic effects. Fumes are very irritating to eyes, lungs and mucous membranes. Inhalation of vapour can cause serious damage to the lungs. Liquid very corrosive to skin, eyes and to the alimentary tract if swallowed.

Incompatibility Reacts explosively with water giving sulphuric acid and fumes of hydrogen chloride. A powerful oxidising agent, it reacts violently with combustible material, e.g. phosphorus, and with silver nitrate.

Handling In fume cupboard. Wear gloves and face shield.

Storage Acids store, never in the laboratory.

Disposal Wearing face shield and gloves, pour on to anhydrous sodium carbonate on a large plastic tray set in a dry sink. When reaction is over wash to waste with water.

Spillage Wearing face shield, respirator and gloves open windows. Spread anhydrous sodium carbonate over the spillage. Shovel up and carefully wash to waste. Wash spillage area well.

First Aid

Eyes Irrigate thoroughly with water. Seek medical attention immediately.

Lungs Remove from exposure, rest and keep warm. Seek medical advice.

Mouth Wash well with water. If swallowed give plenty of water to drink followed by milk of magnesia. Seek medical attention.

Skin Wash with large volume of water. Remove contaminated clothing. Seek medical advice.

Local Conditions

CrO_4^{2-} and CrO_7^{2-}

Hazards Very harmful to eyes, lungs, skin, and if swallowed. Corrosive action on skin and mucous membranes. Can cause dermatitis. Long exposure causes skin ulcers. Recognised carcinogens (Sax). *TLV* 0.05 mg m^{-3} as Cr.

Incompatibility Reducing agents and organic materials including wood. Possibility of explosions.

Handling Avoid dust. Wear gloves and eye protection. Work in fume cupboard if much handling to be done.

Storage With oxidising agents.

Disposal Add to water to form a solution and wash to waste with running water.

Spillage Wear gloves, dust respirator and face shield. Wet with water, brush on to shovel and wash to waste. Wash the spillage area thoroughly.

First Aid

Eyes Irrigate with water. Seek medical attention.

Lungs Remove patient from exposure, rest and keep warm. Seek medical advice.

Mouth Wash out mouth and drink water. Seek medical attention.

Skin Wash with water. Remove and wash contaminated clothing. Seek medical advice.

Local Conditions

CrO_3 Red crystalline solid.

Hazards   The dust is corrosive and irritates respiratory system. Both solid and solution cause severe burns to eyes and skin. Very severe irritation and damage if taken into the mouth. Chronic effect: long exposure causes skin ulcers. A recognised carcinogen (Sax). *TLV* 0.05 mg m^{-3} as Cr.

Incompatibility   Very dangerous if in contact with reducing agents or organic matter when violent reaction or explosion possible.

Handling   Great care required. Wear gloves and eye protection, work in fume cupboard.

Storage   With oxidising agents.

Disposal   Add a little at a time to anhydrous sodium carbonate to neutralise and then add mixture to a large quantity of water and wash to waste.

Spillage   Wear gloves and face shield. Spread anhydrous sodium carbonate over spillage and mop up with water. Wash to waste as above. Wash spillage area thoroughly.

First Aid

Eyes   Irrigate with water. Seek medical attention.

Lungs   Remove patient from area. Rest and keep warm. Seek medical advice.

Mouth   Wash out with water and drink water followed by milk of magnesia. Seek medical attention.

Skin   Wash with water. Wash contaminated clothing. Seek medical advice.

Local Conditions

C   Copper compounds

Hazards	Harmful to eyes, lungs and skin. If swallowed, are very toxic causing violent vomiting, diarrhoea and eventual collapse.
Incompatibility	Ethyne (acetylene).
Handling	Wear eye protection and avoid causing dust from compounds. Copper oxides are usually fine powders and must be handled carefully.
Storage	General store.
Disposal	Moisten and add to normal waste bin.
Spillage	Brush on to shovel, moisten and put in waste bin. Wash soluble salts to waste with large volume of water.

First Aid

Eyes	Irrigate with water. Seek medical advice.
Lungs	Remove patient from contamination, rest and keep warm.
Mouth	Wash with water. Drink water. Seek medical advice.
Skin	Wash with water.

Local Conditions

Tar-like liquid.

Hazards Harmful by inhalation, swallowing and skin contact. Flammable. Fire hazard is moderate when exposed to heat or flame. Toxic fumes if heated to decomposition. A recognised carcinogen (Sax).
Flash point 32°C.

Incompatibility Strong oxidising agents.

Handling Wear gloves and eye protection. Use in fume cupboard.

Storage Flammables store.

Disposal Unavoidable discharges, e.g. small quantities such as washings from glassware, etc. should be emulsified and washed to waste.

Spillage Mix with sand, detergent and water and shovel into a bucket. Mop spillage area with detergent and water. The mixture of sand and oil, etc. can then be burned off in an open place.

First Aid

Eyes Irrigate with water. Seek medical advice.

Lungs Remove patient from area of contamination.

Mouth Wash mouth with water. Seek medical advice.

Skin Wash with soap and water.

Local Conditions

Schedule 1 poisons. Crystalline solids. No handling of these cyanides should be done unless an antidote, as described in *First Aid,* and a suitable respirator are available. The latter should be a positive-pressure, self-contained breathing apparatus although canister respirators specifically for hydrogen cyanide could be used.

Hazards Very poisonous. Vapour from these and solids or solutions are poisonous by inhalation, swallowing and skin contact. Early warning of poisoning is by weakness and heaviness of arms and legs, difficulty in breathing, headache, dizziness. These may be followed by pallor, unconsciousness and death.
TLV (skin) 5 mg m^{-3}.

Incompatibility Heat, moisture and acid cause emission of hydrogen cyanide which is extremely toxic and flammable. Carbon dioxide from the air, being acidic, liberates hydrocyanic acid from solids and solutions.

Handling Never in the open laboratory, always in fume cupboard. Wear gloves and eye protection.

Storage Poisons cupboard. Never in open laboratory.

Disposal For large amounts consult Local Authority. For small amounts wear gloves, face shield and respirator. Dissolve in water, add excess of sodium chlorate (I) (sodium hypochlorite, household bleach, or bleaching powder) and allow to react for a period of 24 hours. The cyanide will have been converted to quite harmless cyanate which can be washed down the drain with running water.

Spillage Evacuate the room. Wear respirator, face shield and gloves. For spillage of solutions, scatter bleaching powder on top and mop up leaving the treated cyanide in a bucket for 24 hours before washing to waste. Consult Local Authority if spillage large. Solid cyanides should be carefully brushed up and treated as in disposal.
Alternative neutralising is by iron (II) sulphate which, with the potassium or sodium cyanide, forms the hexacyanoferrate (II) complex.

First Aid

Rescuers must wear respirators whilst giving first aid since there is a risk of vapour inhalation from contaminated clothing. *Obtain medical attention at once.*
If casualty is breathing, break capsule of pentyl nitrite (amyl nitrite) and give to inhale for 15–20 seconds. Repeat every 2 or 3 minutes to a maximum of two capsules. Administer oxygen through a face mask, keep warm and rest.

Eyes Irrigate thoroughly with water. Seek medical attention.

Lungs Remove patient from exposure. Remove all clothing and place in open air. If breathing has stopped apply artificial respiration *other than mouth to mouth, or mouth to nose. Obtain medical attention at once.*

Mouth Make sick by giving cyanide antidote or finger down back of throat. *Obtain medical attention at once.*

Skin Wash with water. Remove all clothing, putting it in open air. Seek medical advice.

Local Conditions

C Cyclohexa-2,5-diene-1,4-dione
Quinone; p-Benzoquinone

$C_6H_4O_2$ Solid, MP 115°C.

Hazards   Solid harmful to skin and eyes. Vapour causes severe irritation to eyes. Odour noticeable at or just above 0.1 ppm.
TLV 0.1 ppm (0.4 mg m^{-3}).

Incompatibility Oxidising agents.

Handling Wear gloves and eye protection. Handle in fume cupboard.

Storage Poisons cupboard. Never in laboratory.

Disposal Dissolve small amounts such as washings of glassware in large excess of water and run to waste.

Spillage Wear gloves and eye protection. Mop up with plenty of water and wash to waste. Ventilate spillage area well.

First Aid

Eyes Irrigate with water. Seek medical attention.

Lungs Remove patient from exposure, rest and keep warm. Seek medical advice.

Mouth Wash thoroughly with water. Seek medical attention.

Skin Wash with soap and water. Remove and wash contaminated clothing. Seek medical advice.

Local Conditions

C_6H_{12} Liquid, BP 81°C.

Hazards   Harmful if vapour inhaled. Liquid is quite harmful to eyes, skin, and if swallowed. Highly flammable.
TLV 300 ppm (1050 mg m^{-3}).
Flash point −17°C.
Autoignition temperature 260°C.

Incompatibility   Dangerous when exposed to heat or flame. Can react with strong oxidising agents.

Handling   Wear gloves and eye protection. In well-ventilated area, but in fume cupboard if handling any more than a few cm³.

Storage   Flammables store.

Disposal   Unavoidable discharges, e.g. small quantities such as washings from glassware, etc. should be emulsified and washed to waste.

Spillage   Evacuate room and turn off all sources of ignition including hot plates. Wearing respirator, eye protection and gloves, add water and detergent and emulsify. Mop up and wash to waste with large quantity of water.
Alternative treatment is to absorb on sand and take in a bucket to open area for evaporation by scattering on the ground.

First Aid

Eyes   Irrigate with water. Seek medical advice.

Lungs   Remove patient from contamination and keep warm.

Mouth   Wash out mouth with water. Seek medical advice.

Skin   Wash with soap and water.

Local Conditions

$C_6H_{11}OH$ Liquid, BP 162°C.

Hazards Vapour and liquid harmful to eyes, lungs, skin, and if swallowed. Narcotic in high concentrations. Flammable. A suspected carcinogen (Sax).
TLV 50 ppm (200 mg m^{-3}).
Flash point 68°C.
Autoignition temperature 340°C.

Incompatibility Strong oxidising agents.

Handling Wear gloves and eye protection and if quantity is large work in fume cupboard.

Storage Flammables store.

Disposal Unavoidable discharges, e.g. small amounts such as washings from glassware, etc. should be emulsified and washed to waste.

Spillage Switch off all sources of ignition, then as for disposal. Can also be absorbed on sand and scattered in open area for evaporation.

First Aid

Eyes Irrigate with water. Seek medical advice.

Lungs Remove patient to fresh air, keep warm.

Mouth Wash mouth with water and drink water. Seek medical advice.

Skin Wash with soap and water.

Local Conditions

C_6H_{10} Liquid, BP 83°C.

Hazards   Vapour irritates eyes, skin and respiratory system. Harmful if liquid swallowed. Highly flammable.
TLV 300 ppm (1015 mg m^{-3}).
Flash point below — 7°C.

Incompatibility   Oxidising agents.

Handling   Wear gloves and eye protection. Keep away from flames, hot plates, ovens, etc. Handle in well-ventilated area.

Storage   Flammables store.

Disposal   Unavoidable discharges, e.g. small quantities such as washings from glassware, etc. should be emulsified and washed to waste.

Spillage   Turn off all sources of ignition. Wearing gloves, respirator and face shield, evacuate the room and treat as for disposal or absorb spillage on dry sand, shovel into bucket and take to open area for evaporation.

First Aid

Eyes   Irrigate with water. Seek medical advice.

Lungs   Remove patient from contamination, rest and keep warm.

Mouth   Wash mouth with water. Seek medical advice.

Skin   Wash with soap and water.

Local Conditions

$CH_3(CH_2)_8CH_3$ Liquid, BP 174°C.

Hazards An asphyxiant, narcotic in high concentration. Slightly harmful by inhalation. Flammable.
TLV not available.
Flash point 46°C.
Autoignition temperature 250°C.

Incompatibility Reacts with oxidising agents.

Handling No special precautions required but keep away from flames.

Storage Flammables store.

Disposal Unavoidable discharges, e.g. small quantities such as washings from glassware, etc. should be emulsified and washed to waste.

Spillage Add water and detergent, mop and wash to waste with large amount of water.

First Aid

Eyes Irrigate with water. Seek medical advice.

Lungs Remove patient from area and keep warm.

Mouth Wash thoroughly with water. Seek medical advice.

Skin Wash with soap and water.

Local Conditions

$ClCO(CH_2)_8COCl$ Liquid.

Hazards	Corrosive. Harmful vapour. Avoid contact with eyes and skin. Poisonous if swallowed. Probably advisable to purchase the 5% solution in tetrachloromethane.
Incompatibility	With water, slight exothermic reaction.
Handling	Wearing gloves and eye protection, handle in a well-ventilated laboratory or in a fume cupboard.
Storage	Poisons cupboard. Never in laboratory.
Disposal	Add small amounts slowly to large volume of water containing detergent and, stirring vigorously, wash to waste with water.
Spillage	Wearing gloves and face shield, treat with large volume of water and detergent, brushing to emulsify. Wash to waste with water.

First Aid

Eyes	Irrigate with water. Seek medical attention.
Lungs	Remove patient from exposure, rest and keep warm. Seek medical advice.
Mouth	Wash with water. Seek medical attention.
Skin	Wash with soap and water. Remove and air clothing outside.

Local Conditions

Solutions of these compounds are usually only prepared as intermediates in the preparation of other products. The salts are water soluble.

Hazards They are explosive in the dry state and in concentrated solution, and no attempt should be made to isolate them from solution. Other associated hazards are due to the toxicity of the nitrite and of the products being prepared. Many azo dyes are suspected carcinogens and no pupil should be permitted to carry out coupling reactions using unknown compounds. If an aromatic amine is incompletely diazotised, any unreacted aromatic amine may couple on to the diazonium salt to yield a 4-aminophenyl-azobenzene. Such compounds are likely to be carcinogenic. Most diazonium salts decompose rapidly on contact with tissue and skin.

Incompatibility Oxidising agents.

Handling Wear gloves and eye protection. Only small scale work desirable. The product should be completely washed away after use.

Storage ——

Disposal Wear gloves and face shield. Add water and warm gently until reaction is complete. Wash to waste with water.

Spillage Wear gloves and face shield. Mop up either with a solution of potassium iodide or with copper (I) chloride in dilute hydrochloric acid. Warm up the suspension when decomposition of the diazonium salts occurs. Wash to waste.

First Aid

Eyes Irrigate with water. Seek medical attention.

Lungs ——

Mouth Wash with water. Seek medical advice.

Skin Wash well with water and then with soap and water.

Local Conditions

$(C_6H_5CO)_2O_2$ White solid.

Must be kept wet. Probably not needed in schools as a safer alternative for catalysing polymerisations, dodecanoyl peroxide (lauroyl peroxide), is available.

Hazards Harmful to eyes, lungs and skin. Poisonous if swallowed. Explosive when dry or when subjected to shock. Flammable.
TLV 5 mg m^{-3}.
Autoignition temperature 80°C.

Incompatibility Various organic and inorganic acids, alcohols, amines, metallic naphthenates, N,N-dimethylphenylamine (N,N-dimethylaniline).

Handling Wear gloves and eye protection. Handle away from flames, hot plates, etc. Treat as explosive substance.

Storage General store. Keep wet in bottle.

Disposal Add small amounts to strong alkali solution and run to waste with running water. *Do not put in waste bin.*

Spillage Moisten with water, shovel into (plastic) bucket and remove to open area. Burn by adding small quantities at a time to a fire.

First Aid

Eyes Irrigate with water. Seek medical attention.

Lungs Remove patient from area of contamination. Keep warm. Seek medical advice.

Mouth Wash with water. Seek medical attention.

Skin Wash well with water. Seek medical advice.

Local Conditions

D Dichlorodimethylsilane

$(CH_3)_2SiCl_2$ Liquid, BP 70.5°C.

Hazards Corrosive liquid. Harmful vapour, irritates eyes, lungs and skin. Highly flammable liquid. Dangerous when heated.
Flash point −9°C.

Incompatibility Dangerous in contact with oxidising agents. Reacts with water or steam to produce heat and give toxic and corrosive fumes.

Handling In fume cupboard, away from flames and heat. Wear eye protection and gloves.

Storage Flammables store.

Disposal Wear gloves and face shield. Absorb on dry sand in a bucket and remove to an open space for evaporation.

Spillage Wear gloves and face shield. Absorb on dry sand, shovel into dry bucket, using a plastic shovel if on rough surface, e.g. concrete. Leave in an open area for evaporation for a few hours. Burning may also be done in open area.

First Aid

Eyes Irrigate with water. Seek medical attention.

Lungs Remove patient from exposure, rest and keep warm. Seek medical advice.

Mouth Wash mouth with water. Give plenty of water to drink. Seek medical attention.

Skin Wash thoroughly with water. Remove any contaminated clothing. Seek medical advice.

Local Conditions

CH_2ClCH_2Cl Liquid, BP 83.5°C.

Hazards Harmful if inhaled, swallowed or absorbed through the skin or eyes, damaging cornea, liver and kidneys. Exposure to low concentrations for prolonged period may cause tremors, leucocytosis and dermatitis. Highly flammable. Dangerous if exposed to heat or flame.
TLV 50 ppm (200 mg m^{-3}).
Flash point 13°C.
Autoignition temperature 413°C.

Incompatibility Oxidising agents.

Handling In fume cupboard. Wear gloves and eye protection.

Storage Flammables store.

Disposal Unavoidable discharges, e.g. small quantities such as washings from glassware, etc. should be emulsified and washed to waste.

Spillage Wear face shield, respirator and gloves. Turn off all sources of ignition. Apply water and detergent, emulsify by brushing. Mop up and wash to waste with running water. Alternatively absorb the liquid on sand and remove to open area for evaporation. Wash area of spillage with water and detergent.

First Aid

Eyes Irrigate with water. Seek medical attention.

Lungs Remove patient from exposure, rest and keep warm. Seek medical advice.

Mouth Wash the mouth with water. Seek medical attention.

Skin Wash the skin with water. Remove contaminated clothing and wash. Seek medical advice.

Local Conditions

CH_2Cl_2 Liquid, BP 40°C.

Hazards Very harmful if inhaled, swallowed or absorbed through the skin. It is very dangerous to the eyes. It has strong narcotic powers. Decomposes at high temperatures giving toxic fumes.
TLV 200 ppm (720 mg m^{-3}).
Autoignition temperature 631°C.

Incompatibility Strong oxidising agents and alkali metals.

Handling In well-ventilated area, wearing gloves and eye protection if exposure is just for a short time, otherwise use fume cupboard.

Storage Flammables store, not because of flammability but because of high volatility.

Disposal Unavoidable discharges, e.g. small quantities such as washings from glassware, etc. should be emulsified and washed to waste.

Spillage Evacuate room. Wear respirator, face shield and gloves. As for disposal, or absorb on dry sand and remove in a bucket to open area for evaporation.

First Aid

Eyes Irrigate with water. Seek medical attention.

Lungs Remove patient from area, rest and keep warm. Seek medical attention.

Mouth Wash out with water. Seek medical advice.

Skin Wash with soap and water. Very volatile and evaporates off rapidly.

Local Conditions

Di(dodecanoyl) peroxide D
Lauroyl peroxide

$(C_{11}H_{23}CO)_2O_2$ White powder, MP 55°C.

Hazards The powder is irritant to eyes, lungs and skin. It can cause burns on the skin and mucous membranes. Powerful oxidising agent like other organic peroxides. Explosive when dry. Less dangerous than di(benzenecarbonyl) peroxide (benzoyl peroxide).

Incompatibility Any reducing agents and combustible materials.

Handling Wear eye protection and gloves. Handle away from flames and heat. Use spatula and avoid raising dust. Treat as explosive substance.

Storage General store with oxidising agents, as cool as possible but always below 27°C. Keep moist in bottle.

Disposal Wearing face shield, add small amounts to strong alkali solution and wash to waste with running water. Alternatively small amounts may be cautiously added to a small fire of wood shavings on open ground.

Spillage Moisten well with water and shovel into bucket. Add water, stir and wash to waste with running water or burn as in disposal.

First Aid

Eyes Irrigate with water. Seek medical attention.

Lungs Remove patient from area, keep warm. Seek medical advice.

Mouth Wash mouth with water. Seek medical advice.

Skin Wash immediately with water. Seek medical advice.

Local Conditions

$(C_2H_5)_2NH$ and solution in water. Liquid, BP 55.5°C.

Hazards The vapour irritates eyes and respiratory system. Liquid irritates eyes and skin. The solution in water is safer to use and store. Reacts with nitrous acid to give the potent carcinogenic N-nitrosamine.
Highly flammable.
TLV 25 ppm (75 mg m³).
Flash point −18°C.
Autoignition temperature 312°C.

Incompatibility Oxidising agents.

Handling In fume cupboard, wearing gloves and eye protection.

Storage Flammables store.

Disposal Dilute by adding to large volume of water, neutralise and run down drain with plenty of water.

Spillage Evacuate room, wear respirator, face shield and gloves. Turn off all sources of ignition. Mop up with water and wash to waste with running water.

First Aid

Eyes Irrigate with water. Seek medical attention.

Lungs Remove patient from area, rest and keep warm. Seek medical advice.

Mouth Wash out with water and drink water. Seek medical attention.

Skin Wash with water. Remove and wash contaminated clothing. Seek medical advice.

Local Conditions

I_2Cl_6 Orange/yellow crystals.

Hazards Vapour of pungent odour which is very harmful to eyes and respiratory system. Solid burns the skin. Severe internal damage if swallowed.

Incompatibility Reacts with water, steam, acid or acid fumes to give highly toxic fumes.

Handling In fume cupboard. Wear gloves and eye protection.

Storage With oxidising agents.

Disposal Mix with sodium carbonate. Add water carefully and finally wash to waste with running water.

Spillage Wear face shield and gloves. Spread sodium carbonate over the spillage and carefully mop up with water. Wash to waste with running water.

First Aid

Eyes Irrigate with water. Seek medical attention.

Lungs Remove patient from area, rest and keep warm. Seek medical advice.

Mouth Wash with water. If swallowed give water to drink followed by milk of magnesia. Seek medical attention.

Skin Wash with water. Burns require medical attention.

Local Conditions

D Dimethylamine

$(CH_3)_2NH$ Gas, BP 7°C. Usually available as aqueous solutions.

Hazards Vapour, liquid and solutions are irritant to eyes, lungs and skin. Poisonous if swallowed. Reacts with nitrous acid to give the potent carcinogenic N-nitrosamine. Vapour, liquid and solutions of high concentration are flammable.
TLV 10 ppm (18 mg m^{-3}).
Flash point −6°C for 25% W/V solution, −16°C for 40% W/V solution, −50°C for gas.
Autoignition temperature 470°C.

Incompatibility Dangerous when exposed to flame or hot surfaces. Can react vigorously with oxidising agents.

Handling Wear gloves and eye protection, handle in fume cupboard.

Storage Flammables store.

Disposal Small quantities can be dissolved in water, neutralised and washed to waste with water.

Spillage Turn off all flames and sources of ignition, e.g. hot plates. Wear respirator, face shield and gloves, mop up with plenty of water and wash to waste with large volume of water.

First Aid

Eyes Irrigate with water. Seek medical attention.

Lungs Remove patient to fresh air, rest and keep warm. Seek medical advice.

Mouth Wash mouth with water. If swallowed give water to drink. Seek medical attention.

Skin Wash with water. Remove contaminated clothing and wash it. Seek medical advice.

Local Conditions

$C_6H_4(CH_3)_2$ Colourless liquids.
1,2- or o-dimethylbenzene, BP 144°C.
1,3- or m-dimethylbenzene, BP 139°C.
1,4- or p-dimethylbenzene, BP 138°C.

Hazards Vapour is harmful to eyes and respiratory system. Liquid is harmful by skin absorption, to eyes, and if swallowed. Some samples contain benzene as an impurity which can cause damage to blood-forming system. Highly flammable.
TLV (skin) 100 ppm (435 mg m^{-3}).
Flash points 17–25°C.
Autoignition temperatures 464–529°C.

Incompatibility Oxidising agents.

Handling Wear gloves and eye protection. Work in well ventilated area, away from flames.

Storage Flammables store. Never in laboratory.

Disposal Wear gloves and face shield. Unavoidable discharges, e.g. small quantities such as washings from glassware, etc. should be emulsified and washed to waste.

Spillage Wear gloves, face shield and respirator. Turn off all sources of ignition. Add water and detergent and emulsify, finally washing to waste with running water. Alternatively, absorb on sand, and remove outdoors for evaporation.

First Aid

Eyes Wash with water. Seek medical attention.

Lungs Remove patient from area, rest and keep warm. Seek medical advice.

Mouth Wash with water. Seek medical advice.

Skin Wash with soap and water. Seek medical advice.

Local Conditions

$C_6H_3(NO_2)_2(OH)$
2,4-dinitrophenol, MP 112°C.
2,3-dinitrophenol, MP 144°C.
2,6-dinitrophenol, MP 63°C.
Yellow crystals. Solutions often used as pH indicators.

Hazards Extremely poisonous by ingestion, by inhalation of dust and by skin absorption. Prolonged exposure to low concentrations very dangerous. Explode when heated. A fire hazard if heated with flammable materials.

Incompatibility Reducing agents.

Handling Wear gloves and eye protection. Keep moist. Handle in well-ventilated laboratory or fume cupboard.

Storage Poisons cupboard. Keep moist with at least 20% by weight of water.

Disposal Wear gloves and face shield. Small amounts such as washings of glassware can be washed to waste with running water. If amounts larger than this, consult Local Authority.

Spillage Ensure that solid is slightly damp and shovel carefully into large beaker. If quantity small, aid dissolution by addition of sodium hydroxide solution and wash to waste. If large, add some water and consult Local Authority.

First Aid

Eyes Irrigate with water. Seek medical attention.

Lungs Remove patient from exposure. Rest and keep warm. Seek medical attention.

Mouth Wash out with water. If any has been swallowed give water to drink and seek medical attention.

Skin Wash thoroughly with water and then with soap and water. If exposure large, seek medical attention immediately.

Local Conditions

$(NO_2)_2C_6H_3NHNH_2$ Red solid, MP 200°C.

Hazards   Dangerous explosive. Harmful by inhalation of dust and by skin absorption.

Incompatibility   Oxidising agents.

Handling   Store moist and use minimum quantity for experiment. Do not dry unless really necessary. Best used in solution by pupils. Wear gloves and eye protection.

Storage   General store. Have minimal quantity in store and keep moist.

Disposal   Wear gloves and face shield. Mix with very moist sand and allow to dry in open area and burn off in small portions.

Spillage   As for disposal.

First Aid

Eyes   Irrigate with water. Seek medical attention.

Lungs   Seek medical advice.

Mouth   Wash with water. Seek medical advice.

Skin   Wash with soap and water.

Local Conditions

Disulphur dichloride
Sulphur monochloride

S_2Cl_2 Yellowish-red oily fuming liquid with a strong choking vapour, BP 138°C.

Hazards Corrosive causing burns. Very harmful if ingested or inhaled. Decomposes on contact with water to give hydrogen chloride, thiosulphuric acid and sulphur.
TLV 1 ppm (6 mg m^{-3}).
Flash point 118°C.
Autoignition temperature 234°C.

Incompatibility Oxidising agents, chromium (VI) dichloride dioxide (chromyl chloride), water, sodium peroxide, phosphorus (III) oxide (phosphorus trioxide), unsaturated compounds, mercury (II) oxide (mercuric oxide). Potassium, antimony and sulphides of some metals.

Handling Wear eye protection and gloves. Work in fume cupboard and use small scale.

Storage Only in small amounts with flammables.

Disposal Wear gloves and face shield. Allow to evaporate from a safe outside place. Alternatively, slowly add small portions to a large volume of water. When reaction is complete, neutralise and run to waste.

Spillage Ventilate and evacuate room. Turn off all sources of ignition. Wear gloves and face shield. Soak into *dry* sand, transfer to safe outdoor site and treat as in disposal. Wash spillage site well.

First Aid

Eyes Wash with lots of water. Seek medical attention.

Lungs Remove patient from area, keep warm and rest. Seek medical attention.

Mouth Wash out well. If swallowed give water to drink. Seek medical attention.

Skin Wash well with water. Seek medical advice.

Local Conditions

CH_3CHO Liquid, BP 20.8°C. Fruity odour.

Hazards Vapour is irritant to eyes, skin and lungs. Liquid, if swallowed, causes severe irritation of stomach, etc. Continued exposure to low concentration of vapour is dangerous. Highly flammable.
TLV 100 ppm (180 mg m^{-3}).
Flash point −38°C.
Autoignition temperature 207°C.

Incompatibility Strong oxidising agents, alkalis, amines, halogens, phenols.

Handling Wear gloves and eye protection. In well-ventilated area and well away from flames, hot plates, night store heaters, etc.

Storage Flammables store.

Disposal Up to 10 cm^3 can be washed to waste with running water since it is miscible with water.

Spillage Evacuate the room. Turn off all sources of ignition; if spillage is large wear respirator, face shield and gloves. Mop up with plenty of water and wash to waste. Ventilate room thoroughly.

First Aid

Eyes Irrigate with water. Seek medical attention.

Lungs Remove patient from area, rest and keep warm. Seek medical advice.

Mouth Wash mouth with water. If swallowed drink water. Seek medical advice.

Skin Wash with water.

Local Conditions

E Ethanal tetramer
Metaldehyde

$(C_2H_4O)_4$ White crystals, MP 246°C.

Hazards	Very harmful by inhalation, swallowing, skin contact. Can cause kidney and liver damage. Flammable. *Flash point* 34°C.
Incompatibility	Strong oxidising agents.
Handling	Wear gloves and eye protection. Work in well-ventilated space or fume cupboard.
Storage	Flammables store.
Disposal	For small amounts wet crystals thoroughly and mix with sand, put in polythene bag in normal waste.
Spillage	As for disposal.

First Aid

Eyes	Irrigate with water. Seek medical attention.
Lungs	Remove patient from contamination. Seek medical advice.
Mouth	Wash out. Seek medical advice.
Skin	Wash with water, then soap and water. Seek medical advice.

Local Conditions

$(CH_3CHO)_3$ Colourless liquid, BP 128°C.

Hazards   Vapour and liquid are very poisonous. Harmful to eyes, skin and clothing. Highly flammable.
Flash point 36°C.
Autoignition temperature 238°C.

Incompatibility   Produces toxic fumes when heated. Reacts with strong oxidising agents.

Handling   In fume cupboard, wear gloves and eye protection.

Storage   Poisons cupboard.

Disposal   Unavoidable discharges, e.g. small quantities such as washings from glassware, etc. should be emulsified and washed to waste.

Spillage   Evacuate the area, turn off all sources of ignition. Wear gloves, face shield and respirator. Emulsify with water and detergent, mop up with sand, place in suitable container and consult Local Authority.

First Aid

Eyes   Irrigate with water. Seek medical attention.

Lungs   Remove patient to fresh air, rest and keep warm. Seek medical advice.

Mouth   Wash out mouth with water. Seek medical attention.

Skin   Wash with soap and water.

Local Conditions

E Ethanedioic acid and ethanedioates
Oxalic acid and oxalates

$(COOH)_2, 2H_2O$ and $C_2O_4^{2-}$

Hazards Dust irritates respiratory system. Dust and solutions irritate the eyes. Very poisonous if swallowed. Prolonged exposure to dust may cause loss of weight, increasing weakness and dermatitis. *TLV* 1 mg m^{-3}.

Incompatibility Mercury and silver salts decompose vigorously on heating.

Handling Wear eye protection and gloves. Avoid contact with hands, wash at once if contaminated with solid or solution.

Storage Poisons cupboard.

Disposal Add calcium chloride solution to precipitate insoluble calcium ethanedioate (oxalate), dilute with water and wash to waste with running water.

Spillage Wear gloves and face shield, add calcium chloride solution, mop up with plenty of water and wash to waste with running water.

First Aid

Eyes Irrigate with water. Seek medical attention.

Lungs Remove patient to fresh air, rest and keep warm. Seek medical advice.

Mouth Wash with water. If swallowed drink water followed by milk of magnesia. Seek medical attention.

Skin Wash with water. Seek medical advice.

Local Conditions

CH$_3$COOH   Liquid/Solid, MP 16°C. Smells of vinegar.

Hazards   The concentrated acid is poisonous if swallowed. Both liquid and vapour are irritating to the skin and eyes and can cause burns and ulcers. Dilute ethanoic acid is relatively harmless. Flammable. Emits toxic fumes when heated to decomposition.
TLV 10 ppm (25 mg m^{-3}).
Flash point 40°C.
Autoignition temperature 427°C.

Incompatibility   Strong oxidising agents, polyhydric alcohols, chromic (VI) acid, ethane-1,2-diol (ethylene glycol), hydroxyl-containing compounds, nitric acid, chloric (VII) acid (perchloric acid), manganates (VII) (permanganates), peroxides.

Handling   Wear gloves and eye protection; handle in fume cupboard if large quantities being used.

Storage   With acids in general store, up to 250 cm^3 in laboratory.

Disposal   Neutralise with sodium carbonate and wash to waste with running water.

Spillage   Evacuate room; turn off all possible sources of ignition; wear face shield, respirator and gloves. Neutralise with sodium carbonate and wash to waste with running water. Ventilate area.

First Aid

Eyes   Irrigate with water. Seek medical attention.

Lungs   Remove patient from exposure; rest and keep warm. Seek medical advice.

Mouth   Wash mouth with water. If swallowed, give water to drink followed by milk of magnesia. Seek medical advice.

Skin   Drench with water; remove contaminated clothing and wash before re-use. Seek medical advice.

Local Conditions

Ethanoic anhydride
Acetic anhydride

$(CH_3CO)_2O$ Liquid, BP 140°C. Lachrymatory.

Hazards Poisonous if swallowed, causing immediate irritation, pain and vomiting. The liquid irritates and may burn the skin severely. The vapour irritates the respiratory system. The vapour irritates and the liquid burns the eyes severely, with delayed damage. Flammable. Heated to decomposition emits toxic fumes.
(C) *TLV* 5 ppm (20 mg m⁻³).
Flash point 49°C.
Autoignition temperature 390°C.

Incompatibility Chromic (VI) acid, chloric (VII) acid (perchloric acid), sodium peroxide, nitric acid. Reacts violently with water (sometimes delayed).

Handling Preferably in fume cupboard. Wear gloves and eye protection.

Storage Flammables store.

Disposal Add very slowly to large volume of water. Allow to decompose, neutralise with sodium carbonate and wash to waste with running water.

Spillage Turn off all sources of ignition. Wearing respirator, face shield and gloves, sprinkle *dry* sand on top, lift into bucket and carry to a safe place preferably out of doors. Add the mixture to a large volume of water and proceed as in disposal. Wash the site of spillage well with water and detergent.

First Aid

Eyes Irrigate with water. Seek medical attention.

Lungs Remove patient from exposure; rest and keep warm. Seek medical advice.

Mouth Wash with water. If swallowed give plenty of water to drink followed by milk of magnesia. Seek medical advice.

Skin Drench with water. Remove contaminated clothing and wash before re-use. Seek medical advice.

Local Conditions

C_2H_5OH Liquid, BP 78°C.

Hazards High concentrations of vapour and concentrated solutions are dangerous. Can be poisonous by skin absorption in large quantities. Highly flammable. The denaturing substances present in industrial spirit and in mineralised methylated spirits increase the toxicity.
TLV 1000 ppm (1900 mg m⁻³).
Flash point 12°C.
Autoignition temperature 423°C.

Incompatibility Reacts vigorously with strong oxidising agents, concentrated acids such as sulphuric and nitric, phosphorus pentachloride, and with alkali and alkaline earth metals.

Handling Well away from flames and hot surfaces. Eye protection advisable.

Storage Flammables store. Up to 250 cm³ in laboratory.

Disposal Dilute with large quantity of water and wash down the drain.

Spillage Turn off all sources of ignition. Mop up with large quantity of water. Finally wash to waste with running water. Ventilate the area thoroughly.

First Aid

Eyes Irrigate with water. Seek medical advice.

Lungs Remove patient to fresh air, rest and keep warm.

Mouth Wash with water, drink water. Seek medical advice.

Skin Wash with water.

Local Conditions

E Ethanoyl chloride
Acetyl chloride

CH_3COCl Liquid, BP 51°C. Lachrymatory.

Hazards Poisonous if swallowed. Liquid and vapour burn skin, eyes and respiratory tract. Highly flammable.
Flash point 5°C.
Autoignition temperature 390°C.

Incompatibility Violent reaction with water and lower alcohols.

Handling Wear gloves and eye protection; use in fume cupboard.

Storage Flammables store.

Disposal For small quantities wash to waste with water. Quantities larger than 50 cm³ treat as in spillage.

Spillage Evacuate area. Turn off all possible sources of ignition. Wearing respirator, face shield and gloves sprinkle *dry* sand on top, lift into bucket and carry to a safe place and slowly add to a large volume of water. Add sodium carbonate, allow to decompose and decant off the solution. Run to waste with a large volume of water. Wash site of spillage with sodium carbonate and then with water.

First Aid

Eyes Irrigate with water. Seek medical attention.

Lungs Remove patient from exposure, rest and keep warm. Seek medical advice.

Mouth Wash out thoroughly with water. If swallowed give plenty of water to drink followed by milk of magnesia. Seek medical advice.

Skin Drench with water. Remove and wash contaminated clothing before re-use.

Local Conditions

$CH_3COOCHCH_2$ Colourless liquid, BP 73°C.

Hazards	Vapour is harmful when in high concentrations. Liquid irritates eyes and skin. Harmful if swallowed. Highly flammable. Heated to decomposition emits acrid fumes. *TLV* 10 ppm (30 mg m^{-3}). *Flash point* −8°C. *Autoignition temperature* 427°C.
Incompatibility	Reacts with strong oxidising agents.
Handling	Wear gloves and eye protection. Handle in fume cupboard away from flames.
Storage	Flammables store. Never in laboratory.
Disposal	Wear gloves and face shield. Unavoidable discharges, e.g. small quantities such as washings from glassware, etc. should be emulsified and washed to waste
Spillage	Wear gloves and face shield. Turn off all sources of ignition. Emulsify with water and detergent and wash to waste with running water. Alternatively absorb on sand and remove to outside for evaporation.

First Aid

Eyes	Irrigate with water. Seek medical advice.
Lungs	Remove patient from area to fresh air, rest and keep warm.
Mouth	Wash with water. Seek medical advice.
Skin	Wash with soap and water.

Local Conditions

E Ethoxyethane
Diethyl ether

$C_2H_5OC_2H_5$ Liquid, BP 35°C.

Hazards Highly flammable. Flash point and autoignition temperature very low with result that even black heat can ignite an air/ether mixture.

Not particularly poisonous but the vapour inhaled or the liquid swallowed can produce intoxication, drowsiness and unconsciousness. Danger of explosion when liquid evaporated to dryness due to the presence of unstable peroxides. Use only peroxide-free ethoxyethane for experiments involving evaporation. (Peroxide can be removed effectively by shaking the liquid with a suitable reducing agent such as 5% aqueous iron (II) sulphate solution. The ethoxyethane should be tested for peroxides.)

Test for peroxides. To a 2–3 cm³ of ether in a test-tube add an equal volume of 2% potassium iodide plus two or three drops of 2M hydrochloric acid, cork lightly, wrap in towel and shake gently. A brown or light brown colouration in the bottom (water layer) indicates iodine and thence the presence of peroxides. Starch can be added to test for iodine if brown colour not very definite.

TLV 400 ppm (1200 mg m^{-3}).
Flash point −45°C.
Autoignition temperature 180°C.

Incompatibility Strong oxidising agents, e.g. nitric acid.

Handling All sources of ignition should be turned off. The dense vapour can travel along a bench and be ignited by a flame or hot surface.

Storage Flammables store. (In an amber-coloured bottle away from direct light.) Replace stock annually.

Disposal Wear face shield and gloves. If sodium wire has been present to dry the ethoxyethane great care must be taken, for example it would ignite if added to water. Add dry propan-2-ol to the liquid until no further effervescence takes place, leave for a day and shake occasionally, a loose cotton wool plug being used as stopper. The mixture can then be spread out in an open area for evaporation. If no sodium present, the liquid can be emulsified with water and detergent, and spread out in open area.

Spillage Turn off all possible sources of ignition. Wear respirator, face shield and gloves. If sodium not present, apply water and detergent, transfer to a bucket and spread in open place for evaporation. If sodium is present, a dry brush should be used and the liquid transferred to a bucket which has a lid. The liquid should then be treated as described in *Disposal.* Ensure no sodium is left on brush.

First Aid

Eyes Irrigate with water. Seek medical attention.

Lungs Remove patient from area, rest and keep warm. Seek medical advice.

Mouth Wash out with water. Seek medical advice.

Skin Evaporation is so rapid, no treatment is necessary.

Local Conditions

$C_2H_5NH_2$ Liquid, BP 17°C. Usually available as aqueous solutions.

Hazards   High toxicity. Vapour very irritant to eyes, lungs and skin. Poisonous if taken by mouth. Highly flammable.
TLV 10 ppm (18 mg m^{-3}).
Flash point -18°C for gas and 22°C for 70% W/V solution.
Autoignition temperature 384°C.

Incompatibility   Reacts with oxidising agents, e.g. chlorates (I) (hypochlorites).

Handling   In well-ventilated area or fume cupboard and well away from flames or hot surfaces. Wear gloves and eye protection.

Storage   Flammables store.

Disposal   Turn off all sources of ignition. Wear gloves and face shield. Add to large amount of water, neutralise and wash to waste with running water.

Spillage   Turn off all sources of ignition. Wear respirator, face shield and gloves. Mop up with plenty of water and proceed as in disposal.

First Aid

Eyes   Irrigate with water. Seek medical attention.

Lungs   Remove patient to fresh air, rest and keep warm. Seek medical advice.

Mouth   Wash with water, drink water. Seek medical attention.

Skin   Wash with water. Remove clothing if contaminated. Seek medical advice.

Local Conditions

$CH_3COOC_2H_5$ Liquid, BP 77°C.

Hazards	Harmful to eyes, lungs and skin. Liquid is harmful if swallowed. Prolonged exposure to low concentrations of vapour can cause corneal cloudiness and anaemia. Highly flammable. *TLV* 400 ppm (1400 mg m⁻³). *Flash point* −4.4°C. *Autoignition temperature* 427°C.
Incompatibility	Reacts with strong oxidising agents.
Handling	With gloves and eye protection, in well-ventilated area.
Storage	Flammables store.
Disposal	Turn off all sources of ignition. Unavoidable discharges, e.g. small quantities such as washings from glassware, etc. should be emulsified and washed to waste. Alternatively soak up on sand and transport to safe outside area for evaporation.
Spillage	As for disposal.

First Aid

Eyes	Irrigate with water. Seek medical attention.
Lungs	Remove patient to fresh air, and keep warm. Seek medical advice.
Mouth	Wash out mouth with water, drink water. Seek medical advice.
Skin	Wash with soap and water.

Local Conditions

Made by mixing together immediately before use.
Fehling's solution No 1 contains copper (II) sulphate $CuSO_4$;
Fehling's solution No 2 contains approximately 4M sodium
hydroxide NaOH and sodium potassium 2,3-
dihydroxybutanedioate (sodium potassium tartrate).

Hazards   Mainly due to corrosive nature of a heated alkaline solution.
Contact with eyes or skin harmful. Safer alternatives include
Benedict's solution and Barfoed's solution (this gives a positive
result for the same reducing sugars as are detected by Fehling's
solution with the exception of lactose and maltose).

Incompatiblity   Strong acids, zinc, aluminium.

Handling   Use only very small amounts and use water bath. Wear eye
protection and gloves.

Disposal   Wear eye protection and gloves. Neutralise and wash to waste
with plenty of water.

Spillage   Wear gloves and face shield. Mop up into plastic bucket, neutralise and wash to waste.

First Aid

Eyes   Irrigate with water. Seek medical attention immediately.

Lungs ——

Mouth   Wash thoroughly with water. Give plenty of water to drink. Seek
medical attention.

Skin   Soak with water. Remove contaminated clothing. Seek medical
advice.

Local Conditions

F^- and HF_2^-

Hazards	Dust is very harmful to eyes, lungs and skin. Poisonous if solid or solution swallowed. Chronic poisoning results in anaemia, anorexia and dental defects. *TLV* as low as 3 ppm (2.5 mg m^{-3}).
Incompatiblity	When heated or in contact with acids or acid vapour highly toxic fumes are produced.
Handling	Avoid producing dust, use spatula carefully. Gloves should be worn when working with solid or solutions.
Storage	Poisons cupboard.
Disposal	Small quantities can be washed to waste with water.
Spillage	Carefully shovel into bucket, seal in bag and consult Local Authority.

First Aid

Eyes	Irrigate with water. Seek medical attention.
Lungs	Remove patient to fresh air, rest and keep warm. Seek medical advice.
Mouth	Wash out mouth. If swallowed drink water. Seek medical attention.
Skin	Wash with water.

Local Conditions

Gases likely to be stored in cylinders in schools include nitrogen, hydrogen and oxygen (permanent gases); sulphur dioxide, carbon dioxide and possibly even chlorine in lecture-size bottles (liquified gases). Ethyne (acetylene) which may possibly be used in some technical departments is in fact dissolved in propanone absorbed on kapok. Some Local Authorities may well prohibit the storage and use of cylinders of certain gases.

Hazards Each gas contained within the cylinder may present its own specific hazards as to toxicity and flammability during use, but one of the biggest dangers is due to mishandling of the cylinder. A cylinder with its neck broken off would become a jet-propelled missile and great care is needed with transportation and use. Ethyne (acetylene) reacts with copper and copper alloys containing more than 50% copper to form explosive carbides (acetylides). (See booklet published by British Oxygen Company entitled *Safety in the use of Compressed Gas Cylinders* which covers the recommended use of cylinders and of acetylene in particular.)

Problems of corrosion can readily occur with chlorine and sulphur dioxide syphons (see sulphur dioxide). Oxygen can cause ignition of grease and no valve or regulator should be oiled or greased.

In the case of reduction experiments or where hydrogen gas is to be heated or the excess burned away, the apparatus should be checked first for complete removal of air prior to lighting any Bunsens, etc. This is best done by collecting test-tube samples at the exit end of the apparatus until a sample tested at a distance burns quietly.

Care should be taken to ensure that the correct cylinder is used. The colour code used in the UK is:

black	oxygen
grey with black neck	nitrogen
black	carbon dioxide
bright red	hydrogen and coal gas
maroon	ethyne (acetylene)

But the correct means of identification is by the name stamped on the cylinder.

Too rapid a release of gas can cause a build-up of static in the control valve sufficient to ignite flammable gases.

Carbon dioxide can cause 'burns' and gloves should be worn (see carbon dioxide). Though not recommended, small camping gas burners (butane canisters) might sometimes be used as temporary substitutes for Bunsens. Apart from the danger due to the flammability and possible leakage from these lightweight containers, many are notoriously unstable. Some means of weighting or extending the base should be used. Such canisters should be changed out of doors or in a well-ventilated place away from sources of ignition.

Oxygen gas can be absorbed in layers of clothing and a source of ignition or small spark can result in flash burning.

In the event of a fire, fire officers should be informed of the position of the cylinders even if they are not in the immediate vicinity of the fire.

Handling
1. Transport by trolley and *do not leave standing without support.* Secure cylinder either with a cylinder trolley or cylinder bench clamp and use valve guard.

2. Fit a regulator valve and if leaks are suspected check with a dilute solution of Teepol before bringing into laboratory. The correct valve key and spanner should be used without resorting to undue force. If leak persists return cylinder to the supplier.

3. There should be no direct connection to glassware without the use of a pressure blowout device and no connection should be made until a suitable controlled rate of flow has been achieved.

4. A non-return valve should be used if compressed air or oxygen is used along with a combustible gas in a 'torch'.

5. Do not completely empty a cylinder but leave a small positive pressure and close valve thus keeping the air from diffusing back into cylinder and possibly forming an explosive mixture. (Especially in the case of flammable gases.)

Storage
Cylinders containing liquified gases should be stored in a vertical position, carbon dioxide cylinders in preparation room and sulphur dioxide syphons on open shelf in preparation room. When not in use, store compressed gas cylinders in a cool, dry, well-ventilated store, possibly outside main building, and far from flammables or other fire risks, i.e. in ground floor room against an outside wall. Ideally the store should be separated from the rest of the building by fire-resisting walls and floors and have light, blow out, exterior walls. Flammable gases should be stored separately from oxygen, and sparkproof lights and switches installed.

Local Conditions

$CH_3(CH_2)_4CH_3$ Colourless liquid, BP 69°C.

Hazards Vapour is quite irritant, has narcotic effect in high concentrations. Highly flammable.
TLV 100 ppm (360 mg m^{-3}).
Flash point −23°C.
Autoignition temperature 260°C.

Incompatibility Strong oxidising agents.

Handling In well-ventilated area, wearing gloves and eye protection.

Storage Flammables store.

Disposal Turn off all sources of ignition. Unavoidable discharges, e.g. small quantities such as washings from glassware, etc. should be emulsified and washed to waste.

Spillage Turn off all sources of ignition. Wearing gloves and face shield, add water and detergent. Mop the mixture to emulsify and wash to waste with running water or spread out in an open area for evaporation.

First Aid

Eyes Irrigate with water. Seek medical advice.

Lungs Remove patient to fresh air, rest and keep warm.

Mouth Wash mouth with water. Seek medical advice.

Skin Wash with detergent and water. Wash contaminated clothing.

Local Conditions

$NH_2(CH_2)_6NH_2$ Solid, MP 42°C.

Hazards Irritant to eyes, mucous membranes, skin. Flammable.

Incompatibility Oxidising agents.

Handling In well-ventilated area. Use gloves and eye protection.

Storage General store with alkalis.

Disposal Dilute with water, neutralise and wash to waste with running water.

Spillage Shovel into bucket, add water to dissolve, neutralise carefully with dilute hydrochloric acid and wash to waste with running water.

First Aid

Eyes Irrigate with water. Seek medical attention.

Lungs Seek medical advice.

Mouth Wash with water. Seek medical advice.

Skin Wash with water, then with detergent and water.

Local Conditions

$CH_2CH(CH_2)_3CH_3$ Liquid, BP 63.5°C.

Hazards   Poisonous liquid and vapour. Harmful to eyes and skin. Highly flammable.
Flash point below −7°C.

Incompatibility   Oxidising agents.

Handling   In well-ventilated area using gloves and eye protection; well away from flames, etc.

Storage   Flammables store.

Disposal   Unavoidable discharges, e.g. small quantities such as washings from glassware, etc. should be emulsified and washed to waste.

Spillage   Turn off all sources of ignition. Add water and detergent. Emulsify, mop up and run to waste with running water. Alternatively soak up on sand and spread out in an open area for evaporation.

First Aid

Eyes   Irrigate with water. Seek medical attention.

Lungs   Remove patient from area. Keep warm.

Mouth   Wash with water. Seek medical advice.

Skin   Wash with soap and water.

Local Conditions

Colourless gas, BP −253°C.
Available as compressed gas in cylinders (red with name on shoulder) or prepared when required.

Hazards Mixtures with air (oxygen) explode readily if ignited, especially inside the restricted space of glassware. This has been a common cause of explosions in school laboratories. The air is probably flushed out more thoroughly from apparatus if supply is from a cylinder. Mishandling of a cylinder of any compressed gas can be very dangerous (see gas cylinders). Highly flammable.
Autoignition temperature 585°C.

Incompatibility Oxygen, air, bromine, fluorine, nitrogen dioxide. Chlorine in the presence of sunlight or ultraviolet light.

Handling Use eye protection, and explosion screens for demonstrations. Good ventilation needed.
(a) If prepared from zinc and dilute sulphuric acid, use plastic apparatus and maintain a steady stream of the gas. Dropping funnels are preferred to thistle funnels as the latter may allow air to be pushed into flask on the addition of more acid.
Only after samples collected at the exit end of the combustion tube have been taken some distance away and shown to burn quietly should the jet be lit or the combustion tube heated.
An alternative pupil scale qualitative reduction is by gently heating a mixture of zinc and calcium hydroxide powders at the bottom of a boiling tube. The metal oxide halfway along the tube is strongly heated.
(b) Fix cylinders securely by bench clamp or on cylinder trolley and use valve guard. Fit regulator valve on left-hand thread. Check for any suspected leaks with a dilute solution of Teepol. Do not use undue force to fit regulator or to close valves. Open valve slowly to avoid build up of static, and a suitable flow should be obtained before attaching the apparatus which should have a special blow-out device fitted. The photochemical reaction of hydrogen and chlorine using gas syringes should *not* be attempted.

Storage Cylinders well-secured in cool, ventilated store away from flammables and fire risks.

Local Conditions

H Hydrogen bromide and hydrobromic acid

HBr   Gas at room temperature. Acid is pale yellow solution.

Hazards   Acid as supplied is up to 60% concentration. Extremely irritant and corrosive vapour. Vapour and acid harmful to eyes, lungs, skin and also if swallowed. Prolonged exposure to low concentrations is dangerous.
TLV 3 ppm (10 mg m^3).

Incompatibility   Oxidising agents.

Handling   In fume cupboard. Wear gloves and eye protection.

Storage   With acids in general store.

Disposal   Dilute with large quantity of water, neutralise with sodium carbonate and wash to waste with running water.

Spillage   Wear gloves, face shield and respirator. Spread soda ash (i.e. anhydrous sodium carbonate) over the spillage and mop up with water. Wash to waste with running water.

First Aid

Eyes   Irrigate with water. Seek medical attention.

Lungs   Remove patient from exposure, rest and keep warm. Seek medical advice.

Mouth   Wash with water. If swallowed drink water and follow with milk of magnesia. Seek medical attention.

Skin   Wash with water. Treat with paste of glycerol and magnesia. Wash contaminated clothing. Seek medical advice.

Local Conditions

HCl   Colourless gas. Acid is colourless solution.

Hazards   Irritant vapour harmful to eyes, lungs and skin. The acid burns eyes and skin. If concentrated acid is swallowed it is very irritant and corrosive.
(C) *TLV* 5 ppm (7 mg m^{-3}).

Incompatibility   Oxidising agents. With methanal (formaldehyde) vapour in air it forms chloromethoxychloromethane (bis-chloromethyl ether), a very potent carcinogen with a *TLV* of 0.001 ppm.

Handling   Wear gloves and eye protection.

Storage   With acids in general store.

Disposal   Dilute with large volume of water, neutralise with sodium carbonate and wash to waste with running water.

Spillage   Wear gloves, face shield and respirator. Spread soda ash (i.e. anhydrous sodium carbonate) over the spillage, then add water and mop up, wash to waste with running water.

First Aid

Eyes   Irrigate with water. Seek medical attention.

Lungs   Remove patient from exposure, rest and keep warm. Seek medical advice.

Mouth   Wash with water. Drink water followed by milk of magnesia. Seek medical attention.

Skin   Wash with water. Treat with mixture of glycerol and magnesia. Seek medical advice.

Local Conditions

HF   Colourless gas. Acid is colourless solution.

Hazards   Extremely irritant and harmful vapour, affecting all parts of the respiratory system. Vapour or acid burns eyes and skin. If swallowed causes severe internal damage. Even very slight contact is dangerous.
TLV 3 ppm (2 mg m^{-3}).

Incompatibility   Oxidising agents.

Handling   Maximum care is required. It is essential to wear gloves and face shield and handle in fume cupboard. Even dilute acid is very dangerous.

Storage   Within a polythene vessel in general store with acids, away from glass which becomes etched.

Disposal   Add to large volume of sodium carbonate solution and wash to waste with running water.

Spillage   Wear gloves, face shield and respirator. Spread soda ash (i.e. anhydrous sodium carbonate) over the spillage and mop up very cautiously with plenty of water. Wash to waste with running water.

First Aid

Eyes   Irrigate *immediately* with water and *obtain medical attention at once.*

Lungs   Remove patient from area, rest and keep warm. Obtain medical attention if exposure is other than very slight.

Mouth   Wash out mouth with water. Give water to drink. Obtain medical attention.

Skin   Wash continuously with water until medical attention is obtained.

Local Conditions

HI   Gas. Acid is yellow to brown solution.

Hazards Very irritant vapour, harmful to eyes, lungs and skin. Solution burns eyes and skin and if swallowed is very irritant and corrosive poison.
TLV not available.

Incompatibility Oxidising agents.

Handling In fume cupboard, Wear gloves and eye protection.

Storage Acids store. Not near nitric or sulphuric acid.

Disposal Dilute, neutralise with sodium carbonate and wash to waste with running water.

Spillage Wear gloves, face shield and respirator. Spread soda ash (i.e. anhydrous sodium carbonate) over the spillage then add water and mop up, wash to waste with running water.

First Aid

Eyes Irrigate with water. Seek medical attention.

Lungs Remove patient from exposure, rest and keep warm. Seek medical advice.

Mouth Wash with water. Drink water followed by milk of magnesia. Seek medical attention.

Skin Wash with water and treat with mixture of glycerol and magnesia. Seek medical advice.

Local Conditions

H_2O_2 35% solution (100 vol), BP 107°C.

Hazards At high concentrations the solutions are corrosive, causing burns to eyes, lungs, mouth and skin. In the stomach the sudden evolution of oxygen can cause injury by acute swelling of stomach, producing nausea, vomiting and bleeding.
TLV 1 ppm (1.4 mg m^{-3}).

Incompatibility Acids, flammable materials, reducing agents. Rapid evolution of oxygen if mixed with potassium manganate (VII) (permanganate), manganese (IV) oxide and several other transition metal compounds, and some biological materials containing catalase.

Handling Great care is required with concentrated solutions particularly from 20 volume upwards, i.e. from 6% weight/volume up to 35% weight/volume. Wear gloves and eye protection. Use in well-ventilated area, well away from incompatible materials. On receiving bottle from supplier loosen screw cap slightly. It is economically worthwhile to purchase 100 volume hydrogen peroxide, but upon receipt this should be diluted five times and stored in bottles with loosely screwed-on tops.

Storage In cool dark place away from reducing agents and transition metal compounds (likely catalysts).

Disposal Add to water and wash to waste with running water.

Spillage Wear gloves and face shield, mop up and wash to waste with running water.

First Aid

Eyes Irrigate with water. Seek medical attention.

Lungs ——

Mouth Wash out with water and give more water to drink. Seek medical advice.

Skin Wash with water. Wash contaminated clothing. Seek advice if blisters form.

Local Conditions

H$_2$S   Evil-smelling gas.

Hazards   Extremely poisonous gas. High concentrations cause immediate unconsciousness. Even low concentrations cause irritation of respiratory system and eyes, causing headaches and weakness. An added danger is that after exposure for some time the sense of smell becomes impaired. Flammable, forming explosive mixture with air.
TLV 10 ppm (15 mg m^3).

Incompatibility   Oxidising agents.

Handling   Always in fume cupboard.

Storage   Prepare only when required.

Disposal   Preferably through a water-fed scrubber in the fume cupboard. Reactants from preparation should be left in fume cupboard until no further gas evolved. Solution can then be diluted with large quantity of water and put down the drain.

Spillage   This could be in form of gas escape from apparatus or cylinder, and treatment would be simply to evacuate laboratory and possible adjoining rooms and ventilate thoroughly.

First Aid

Eyes   Irrigate with water. Seek medical advice.

Lungs   Remove patient to fresh air, rest and keep warm. Seek medical advice.

Mouth   Seek medical advice.

Skin ——

Local Conditions

H 2-Hydroxybenzoic acid
Salicylic acid

$(HO)C_6H_4COOH$ White solid, 159°C.

Hazards   Harmful by ingestion, causes nausea, vomiting, etc. Slight fire hazard.
Flash point 157°C.
Autoignition temperature 545°C.

Incompatibility   Strong oxidising agents.

Handling   Eye protection advisable.

Storage   General store. Never in laboratory.

Disposal   Dampen with water, seal in bags and put into waste bin.

Spillage   As for disposal but mix with moist sand before putting in bin.

First Aid

Eyes   Irrigate with water. Seek medical advice.

Lungs   Remove patient from area.

Mouth   Wash out with water. Seek medical advice.

Skin   Only slightly affected by contact. Wash with soap and water.

Local Conditions

I_2 Black solid – sublimes to purple vapour.

Hazards   Vapour is harmful to respiratory system. Vapour and solid irritate the eyes. Solid burns the skin. If swallowed internal irritation is severe. Prolonged exposure to low concentrations of vapour or contact with skin or eyes is dangerous.
(C) *TLV* 0.1 ppm (1 mg m⁻³).
When heated gives large amount of iodine vapour.

Incompatibility   Reducing agents, ethyne, ethanol, ammonia, phosphorus.

Handling   In well-ventilated area. Gloves and eye protection should be worn.

Storage   General store.

Disposal   Dissolve in sodium thiosulphate solution and wash to waste with running water.

Spillage   If large wear gloves, face shield and respirator. Sweep up, add to sodium thiosulphate solution and wash to waste with running water.

First Aid

Eyes   Irrigate with water. Seek medical attention.

Lungs   Remove patient from area, rest and keep warm. Seek medical advice.

Mouth   Wash with water. If swallowed give water to drink. Seek medical attention.

Skin   Swab with 1% solution of sodium thiosulphate.

Local Conditions

1- Iodobutane
Butyl iodide

C_4H_9I Liquid, BP 131°C.

Hazards Low toxicity by inhalation. Harmful to eyes, skin and if swallowed. Flammable.
TLV not available.
Flash point 36°C.

Incompatibility Oxidising agents. Can react with water or steam to produce toxic fumes.

Handling In well-ventilated area. Wear gloves and eye protection.

Storage Flammables store.

Disposal Unavoidable discharges, e.g. small quantities such as washings from glassware, etc. should be emulsified and washed to waste.

Spillage Add water and detergent and emulsify as for disposal but if quantity is larger than about 50 cm³, absorb on sand and transfer to an open area for evaporation.

First Aid

Eyes Irrigate with water. Seek medical attention.

Lungs Remove patient from area, rest and keep warm. Seek medical advice.

Mouth Wash with water, drink water. Seek medical advice.

Skin Wash with soap and water.

Local Conditions

C_2H_5I Colourless to brown liquid, BP 72°C.

Hazards   Moderately toxic by inhalation. Harmful to eyes, skin and if swallowed. Flammable. Chronic effects by inhalation and skin absorption.
TLV not available.
Flash point 71°C.

Incompatibility   Oxidising agents. Can react with water or steam to produce toxic and corrosive fumes.

Handling   In well-ventilated area, with gloves and eye protection.

Storage   Flammables store.

Disposal   Unavoidable discharges, e.g. small quantities such as washings from glassware, etc. should be emulsified and washed to waste.

Spillage   Add water and detergent and emulsify as for disposal but if quantity is larger than about 50 cm³, absorb on sand and transfer to an open area for evaporation.

First Aid

Eyes   Irrigate with water. Seek medical advice.

Lungs   Remove patient from area, rest and keep warm. Seek medical advice.

Mouth   Wash with water. Drink water. Seek medical advice.

Skin   Wash with soap and water.

Local Conditions

I 1-Iodopropane
Propyl iodide

C_3H_7I Liquid, BP 103°C.

Hazards Low toxicity by inhalation. Harmful to eyes, skin, and if swallowed. **Highly flammable.**
TLV not available.
Flash point 22°C.

Incompatibility Oxidising agents. Can react with water and steam to produce toxic fumes.

Handling In well-ventilated area, with gloves and eye protection.

Storage Flammables store.

Disposal Unavoidable discharges, e.g. small quantities such as washings from glassware, etc. should be emulsified and washed to waste.

Spillage Wear gloves and face shield. Add water and detergent and emulsify as for disposal but if quantity is larger than about 50 cm³ absorb on sand and transfer to an open area for evaporation.

First Aid

Eyes Irrigate with water. Seek medical advice.

Lungs Remove patient from area, rest and keep warm.

Mouth Wash with water. Drink water. Seek medical advice.

Skin Wash with soap and water.

Local Conditions

Hazards	Dust or fume inhaled or material swallowed as solid or in solution is very harmful. Cumulative poisons. *TLV* 0.15 mg m^{-3}.
Incompatibility	With hydrogen azide (hydrazoic acid), or hydrazine and nitrous acid (nitric III acid), may form explosive azides.
Handling	Avoid long contact with skin. Electrolysis of molten lead (II) bromide to be carried out in fume cupboard.
Storage	Poisons cupboard.
Disposal	If soluble, small amounts can be dissolved in water and run to waste. For large amounts consult Local Authority.
Spillage	Brush into suitable container, seal and put in waste bin.

First Aid

Eyes	Irrigate with water and seek medical advice.
Lungs	Remove from source of exposure. Seek medical advice.
Mouth	Wash with water. If swallowed give more water or milk to drink. Seek medical advice.
Skin	Wash with soap and water.

Local Conditions

Grey, softish metal.

Hazards Very similar to potassium and sodium. Stored in liquid paraffin since it reacts with oxygen and water but less violently than sodium. Reacts with moisture of skin to give lithium hydroxide which is corrosive. Compounds are harmful by ingestion, inhalation and skin contact. Metal may explode when heated on porcelain.

Incompatibility Carbon dioxide, water, oxidising agents, halogenated hydrocarbons.

Handling Never touch the metal: use gloves, eye protection and tongs. Do not expose dry metal to air for more than a few seconds. If heating metal in air, use spatula or deflagrating spoon rather than porcelain lid.

Storage General store with reducing agents, immersed in liquid paraffin.

Disposal Small quantities up to 1 g by reacting small pieces with water.

Spillage Wear gloves and face shield. Cover with dry sand, shovel into dry container and transport to safe open area. Add in small quantities to ethanol. When reaction is complete run to waste with excess running water.

First Aid

Eyes Irrigate with water. Seek medical attention.

Lungs Seek medical advice.

Mouth Wash with water. Seek medical advice.

Skin Drench with water after removing adhering metal.

Local Conditions

Most are deliquescent.

Hazards Moderately toxic when absorbed by ingestion or by inhalation. Large doses can be fatal and a dose of 3 g produces the symptoms of nausea, ataxia. Prolonged exposure may damage kidneys, but elimination from body is rapid.

Incompatibility Depends on anion present.

Handling Normal.

Storage General store.

Disposal Wash to waste.

Spillage Wear gloves, mop up and wash to waste.

First Aid

Eyes Irrigate with water. Seek medical advice.

Lungs Remove from exposure. Seek medical advice.

Mouth Wash out well with water. If any has been swallowed give more water to drink and seek medical advice.

Skin Wash with water.

Local Conditions

LiH   Grey solid.

Hazards Harmful if inhaled or swallowed. Large doses cause dizziness and loss of consciousness. Harmful to skin, forms lithium hydroxide in contact with moisture.
Ignites spontaneously in moist air, the hydrogen formed burning. Must therefore be stored in dry conditions.
TLV 0.025 mg m^{-3}.

Incompatibility Water and moist substances, and oxidising agents or acids. Violent reactions can occur with strong oxidising agents.

Storage General store with reducing agents – in dry, airproof tin.

Handling In fume cupboard, dry conditions. Wear gloves and eye protection.

Disposal Add the solid in small amounts at a time to a large volume of water. Neutralise and wash to waste with running water.

Spillage Wear gloves and face shield. Mix with dry sand, shovel into dry container, remove to open space and add a little at a time to large volume of water. Wash to waste with running water.

First Aid

Eyes Irrigate with water. Seek medical attention.

Lungs Remove patient from exposure, rest and keep warm. Seek medical advice.

Mouth Wash with water. Give plenty of water to drink followed by 1% ethanoic acid or vinegar. Seek medical advice.

Skin Drench with water. Remove contaminated clothing and wash before re-use.

Local Conditions

LiOH   White crystals.

Hazards	Solid and solution harmful to eyes and burn the skin. Harmful if swallowed.
Incompatibility	Concentrated acids.
Handling	Wear gloves and eye protection.
Storage	General store, with alkalis.
Disposal	Dissolve in water, neutralise and wash to waste with running water.
Spillage	Wearing gloves and face shield, brush on to shovel, dissolve in water, neutralise and wash to waste with running water.

First Aid

Eyes	Irrigate with water. Seek medical attention.
Lungs	——
Mouth	Wash with water and if swallowed drink water. Seek medical advice.
Skin	Wash thoroughly with water.

Local Conditions

Silvery metal.

Hazards Inhalation of fumes of freshly sublimed magnesium oxide may cause metal fume fever. Particles of metallic magnesium or alloy entering the skin may produce a local lesion which is slow to heal. Fire is main hazard. Very dangerous especially in form of powder or flakes. Can be ignited by a spark. May ignite spontaneously when the powder is damp particularly with water oil emulsion. Burns from burning magnesium are very severe. Dry sand is the only means of extinguishing magnesium fires – *never* use water, tetrachloromethane or BCF as explosion will occur.
Magnesium oxide fume has *TLV* of 10 mg m^{-3}.

Incompatibility Moisture, acids, alcohols, oxidising agents, sulphur, sulphates, phosphates (V), nitrates, carbonates, halogenated alkanes, 'teflon', metal oxides, moist silicon dioxide. Reactions are violent if heated and if metal is finely divided.

Handling Wear eye protection and keep well away from flames. Replace stopper as soon as sample of powder removed from bottle. Wear leather gloves if possible.

Storage General store, with reducing agents.

Disposal Add to very dilute hydrochloric acid in well-ventilated space and wash to waste with water.

Spillage Wear gloves and face shield. As for disposal but if quantity is large brush into suitable container and transport to open space for addition to acid.

First Aid

Eyes Irrigate with water. Seek medical attention.

Lungs Remove patient from area. Seek medical advice.

Mouth Wash out.

Skin Wash and remove any particles lodged in skin.

Local Conditions

Magneson I $O_2NC_6H_4NNC_6H_3(OH)_2$
4[(4-nitrophenyl)azo] benzene-1,3-diol
(p-nitrobenzeneazoresorcinol)
Magneson II $O_2NC_6H_4NNC_{10}H_6(OH)$
4[(4-nitrophenyl)azo] naphthalen-1-ol
(p-nitrobenzeneazo-α-naphthol)
Reddish brown solid practically insoluble in water. The reagent
is usually in the form of a solution of 0.001 g of the dyestuff in
100 cm³ of molar sodium hydroxide.

Hazards Due to the corrosiveness of the sodium hydroxide, especially to
eyes, and possible carcinogenic nature of the dyes.

Incompatibility Oxidising agents. Solution with zinc, aluminium.

Handling Wear eye protection and gloves. Should be used on small scale
and only in advanced laboratories or under close supervision.

Storage General store.

Disposal Wear eye protection and gloves. Wash to waste with water.

Spillage Wear eye protection and gloves. Mop up and wash to waste with
water.

First Aid

Eyes Wash with water. Seek medical attention.

Lungs ——

Mouth Wash out with water. Seek medical advice.

Skin Wash with water and then with soap and water.

Local Conditions

M Mercury

Silvery liquid, BP 356.9°C.

Hazards Poisonous vapour and liquid. Continuous exposure to small concentrations of vapour are harmful. Frequent contact with the skin is harmful as a result of skin absorption.
TLV 0.05 mg m^{-3}.

Incompatibility Ethyne (acetylene), ammonia, hydrogen, ethanedioic acid (oxalic acid), bromine. Corrodes metals by amalgamation.

Handling In a well-ventilated area. Wear gloves. Use a large plastic or sealed wooden mercury spillage tray containing water.

Storage General store in locked cupboard. Set polythene containers of mercury in beaker as spillage container.

Disposal Mercury liquid should be cleaned for re-use. Distillation should not be attempted. Cleaning can either be done with Sellotape or dilute nitric acid. Allowance is given against fresh supplies by many firms.

Spillage Any spillage however small should be be dealt with. The mercury can be sucked up by using a plastic syringe. Inaccessible drops in cracks in floor or bench can be covered with sulphur/lime paste in water. Leave for a few hours and sweep into a container for disposal.

First Aid

Eyes Wash with water. Seek medical advice.

Lungs Remove patient from area. Seek medical advice.

Mouth If swallowed obtain medical attention immediately.

Skin Medical observation necessary if skin contact has been prolonged.

Local Conditions

Many of these are Schedule 1 poisons.

Hazards Mercury (II) compounds are generally more toxic than the mercury (I) compounds. Prolonged exposure to low concentrations is dangerous.
Vapour, dust, solutions and solids are all very harmful by inhalation, swallowing, skin contact and to the eyes. Organic mercury compounds readily absorbed through the skin. Alkyl mercurials much more toxic than phenyl mercury compounds. May be encountered as fungicides in seed dressings, swimming pools, etc. Compounds may emit mercury fumes when heated.
TLV 0.05 mg m^{-3}.

Incompatibility Ethanol, hydrocarbons, ethyne (acetylene), ethanedioic acid (oxalic acid).

Handling Wear gloves and eye protection and work in well-ventilated area. Prolonged handling should be done in fume cupboard.

Storage Poisons cupboard.

Disposal Small amounts can be washed to waste with a large volume of water. Usual recommended dilution for disposal 10 ppm mercury.

Spillage Wear gloves and eye protection and collect in suitable container. Consult Local Authority for disposal.

First Aid

Eyes Irrigate with water. Seek medical advice.

Lungs Remove patient from area, rest and keep warm. Seek medical advice.

Mouth Wash mouth with water. If swallowed give large amount of milk to drink. Seek medical attention.

Skin Wash with soap and water. Seek medical advice.

Local Conditions

M Methanal
Formaldehyde; Formalin

HCHO   Gas, but supplied in aqueous solution.

Hazards The vapour is very irritant to eyes and lungs. The solution (formalin) is very irritant to the skin. Solution poisonous if swallowed. Vapour and concentrated solutions are flammable. Explosion hazard when aqueous solutions heated above their flash points. Prolonged exposure can cause hypersensitivity, damage to lungs, and cracking of skin. A suspected carcinogen of the lung (Sax).
(C) *TLV* 2 ppm (3 mg m^{-3}).
Flash point 50°C.
Autoignition temperature 470°C.

Incompatibility Strong oxidising agents. With hydrogen chloride it reacts in air to form chloromethoxychloromethane (bischloromethyl ether), a strong carcinogen, which has *TLV* of 0.001 ppm.

Handling In well-ventilated area for very small quantities but large quantities in fume cupboard. Wear gloves and eye protection and keep well away from flames and hot surfaces. Care should be taken to exclude hydrogen chloride vapour from areas where biological specimens preserved in 'formalin' are stored or examined. Possible sources of hydrogen chloride in a biology laboratory are its use as a component of several chromatographic solvents, as a solvent for some stains, e.g. phloroglucinol (benzene-1,3,5-triol) and in mammalian gut. Alternative fixatives have recently become available.

Storage Flammables store. Do not store with hydrochloric acid or with chlorides which are hydrolysed by moist air.

Disposal Wash to waste with running water.

Spillage Turn off all sources of ignition. Wear gloves, face shield and respirator. Mop up with plenty of water and wash to waste. Ventilate the contaminated area thoroughly.

First Aid

Eyes Irrigate with water. Seek medical attention.

Lungs Remove patient from area, rest and keep warm. Seek medical advice.

Mouth Wash mouth with water. Drink milk. Seek medical attention.

Skin Wash with water. Wash contaminated clothing.

Local Conditions

M Methanoic acid
Formic acid

HCOOH   Liquid 98/100%, BP 100.5°C.

Hazards Vapour very irritant to eyes and lungs. Liquid burns eyes and skin. Very irritant poison if swallowed. Flammable.
TLV 5 ppm (9 mg m⁻³).
Flash point 69°C.
Autoignition temperature 601°C.

Incompatibility Alkalis, oxidising agents.

Handling Wear gloves and eye protection. Can work with small quantities in well-ventilated area but better to use fume cupboard.

Storage With acids in general store.

Disposal Neutralise, dilute with water and wash to waste.

Spillage Evacuate area, wear gloves, face shield and respirator. Cover with sodium hydrogencarbonate, add water and mop up. Wash to waste with running water.

First Aid

Eyes Irrigate with water. Seek medical attention.

Lungs Remove patient from area. Rest and keep warm. Seek medical advice.

Mouth Wash with water. Drink water and take milk of magnesia. Seek medical attention.

Skin Wash with water. Seek medical advice.

Local Conditions

CH_3OH Colourless liquid, BP 65°C.

Hazards Vapour and liquid are harmful to eyes, lungs, skin and other organs, and if swallowed, the effects sometimes appearing several hours later. Prolonged exposure to low concentrations can cause serious illness. Cumulative poison. Highly flammable.
TLV (skin) 200 ppm (260 mg m^{-3}).
Flash point 10°C.
Autoignition temperature 464°C.

Incompatibility Oxidising agents, e.g. bromine, hydrogen peroxide, chlorates (I) (hypochlorites). Phosphorus pentachloride, alkali metals.

Handling In well-ventilated area normally but in fume cupboard for large quantities or prolonged working with the compound. Eye protection advisable.

Storage Flammables store. Up to 250 cm³ in laboratory.

Disposal Dilute with water and wash with running water.

Spillage Evacuate the room. Turn off all sources of ignition. Wearing gloves and respirator, mop up with plenty of water and wash to waste with running water. Ventilate area thoroughly.

First Aid

Eyes Irrigate with water. Seek medical advice.

Lungs Remove patient from exposure, rest and keep warm. If exposure severe seek medical advice.

Mouth Wash mouth with water. Drink water. Seek medical advice.

Skin Wash with water. Remove and wash contaminated clothing.

Local Conditions

CH_3NH_2 Colourless gas, fishy smell. Usually available as aqueous solution.

Hazards   The unpleasant fishy smell is a deterrent to inhaling the gas which is very irritant. The gas also irritates the skin and eyes. The solution in water irritates the skin and it is harmful if swallowed. Gas is highly flammable.
TLV   10 ppm (12 mg m^{-3}).
Flash point $-18°C$ for gas and 7.5°C for 35% W/V solution.
Autoignition temperature   430°C.

Incompatibility   Reacts with oxidising agents.

Handling   In fume cupboard. Wear gloves and eye protection.

Storage   Flammables store.

Disposal   Neutralise with hydrochloric acid in fume cupboard. Solution can then be washed to waste with water.

Spillage   Evacuate the room. Wear gloves, face shield and respirator. Switch off all sources of ignition. If gas only involved, ventilate room thoroughly. If in solution neutralise with dilute hydrochloric acid, mop up with water and wash to waste with running water.

First Aid

Eyes   Irrigate with water. Seek medical attention.

Lungs   Remove patient from exposure, rest and keep warm. If exposure large seek medical advice.

Mouth   Wash out mouth and give water to drink. Seek medical advice.

Skin   Wash with water. Remove and wash contaminated clothing.

Local Conditions

Mineralised. This is a mixture of approximately 90% ethanol, 10% wood spirit which is mainly methanol, and a small amount of paraffin and dye. Highly flammable.

Hazards See ethanol and methanol.

Incompatibility Phosphorus pentachloride, alkali metals, oxidising agents, and concentrated acids such as sulphuric and nitric.

Handling Well away from flames and hot surfaces. Eye protection advisable.

Storage Flammables store. Up to 250 cm³ in laboratory.

Disposal Dilute with water and wash to waste with running water.

Spillage Turn off all sources of ignition, wear gloves and mop up with large quantity of water. Finally wash to waste with running water. Ventilate the area thoroughly.

First Aid

Eyes Irrigate with water. Seek medical advice.

Lungs Remove patient from exposure. Rest and keep warm. If exposure severe seek medical advice.

Mouth Wash mouth with water. Drink water. Seek medical advice.

Skin Wash with water. Remove and wash contaminated clothing.

Local Conditions

$C_6H_5CH_3$ Colourless liquid, BP 111°C.

Hazards   Harmful vapour causing dizziness, headache and nausea. Vapour and liquid irritate eyes and respiratory system. Poisonous by skin absorption. Chronic effects include anaemia and dermatitis. Highly flammable.
TLV (skin) 100 ppm (375 mg m^{-3}).
Flash point 4.4°C.
Autoignition temperature 536°C.

Incompatibility   Reacts vigorously with strong oxidising agents.

Handling   Wear gloves and eye protection. In fume cupboard, away from sources of ignition such as flames and hot plates.

Storage   Flammables store. Never in laboratory.

Disposal   Unavoidable discharges, e.g. small quantities such as washings from glassware, etc. should be emulsified and washed to waste.

Spillage   Wear gloves, face shield and respirator. Turn off all sources of ignition. Add water and detergent, emulsify and wash to waste with running water. Alternatively sand can be added to absorb the liquid and the mixture spread out in an open area for evaporation.

First Aid

Eyes   Irrigate with water. Seek medical advice.

Lungs   Remove patient from area, rest and keep warm. Seek medical advice.

Mouth   Wash with water. Seek medical advice.

Skin   Wash with soap and water. Seek medical advice.

Local Conditions

$C_5H_{11}OH$ Liquid, BP 132°C.

Hazards Vapour is irritant to eyes, lungs, skin. Harmful if swallowed, causes headache, vomiting, etc. Flammable. More toxic than ethanol, but only slowly absorbed. Long exposure can be dangerous.
TLV 100 ppm (360 mg m^{-3}).
Flash point 43°C.
Autoignition temperature 343°C.

Incompatibility Can react vigorously with oxidising agents.

Handling Wearing gloves and eye protection, use in well-ventilated area and well away from open flames, hot plates, etc.

Storage Flammables store. Up to 250 cm³ in laboratory.

Disposal Unavoidable discharges, e.g. small quantities such as washings from glassware, etc. should be emulsified and washed to waste.

Spillage Turn off all sources of ignition. Wearing gloves and face shield, apply water and detergent, brush to form emulsion and wash to waste with water.

First Aid

Eyes Irrigate with water. Seek medical advice.

Lungs Remove patient from area, rest and keep warm.

Mouth Wash with water. If liquid swallowed give water and milk to drink. Seek medical advice.

Skin Wash with water.

Local Conditions

M Methyl ethanoate
Methyl acetate

CH_3COOCH_3 Colourless liquid, BP 58°C.

Hazards   Harmful to eyes, lungs, skin, and if swallowed. Highly flammable.
TLV 200 ppm (610 mg m^{-3}).
Flash point −9°C.
Autoignition temperature 502°C.

Incompatibility   Oxidising agents and lithium aluminium hydride.

Handling   In well-ventilated area well away from flames and other sources of ignition, wearing gloves and eye protection.

Storage   Flammables store. Up to 250 cm³ in laboratory.

Disposal   Dilute with water and wash to waste with running water.

Spillage   Turn off all sources of ignition. Wear gloves and face shield. Mop up with plenty of water and put down drain with large amount of running water.

First Aid

Eyes   Irrigate with water. Seek medical advice.

Lungs   Remove patient from area, rest and keep warm. Seek medical advice if exposure severe.

Mouth   Wash with water, give water to drink. Seek medical advice.

Skin   Wash with water. Wash contaminated clothing before re-use.

Local Conditions

$CH_2C(CH_3)COOCH_3$ Monomer is a colourless liquid, BP 101°C.

Hazards Vapour irritates the eyes and respiratory system. Liquid is irritant to eyes, skin and alimentary system if swallowed. Highly flammable.
TLV (skin) 100 ppm (410 mg m⁻³).
Flash point 10°C.
Autoignition temperature 421°C.

Incompatibility Oxidising agents, even air can cause explosion.

Handling Wear gloves and eye protection, handle in well-ventilated area.

Storage Flammables store.

Disposal Unavoidable discharges, e.g. small quantities such as washings from glassware, etc. should be emulsified and washed to waste.

Spillage Wear gloves and face shield. Turn off all sources of ignition. Add water and detergent. Mop to emulsion and wash to waste with running water. Alternatively soak up on sand and move to safe outdoor area for evaporation.

First Aid

Eyes Irrigate with water. Seek medical advice.

Lungs Remove patient from exposure, rest and keep warm.

Mouth Wash with water. Seek medical advice.

Skin Wash with soap and water.

Local Conditions

$C_6H_4(OH)CH_3$ Solids, MPs 11–35.5°C.

Hazards Harmful vapours, irritate nose, etc. corrosive to mucous membranes. Poisonous by swallowing, skin absorption or by inhalation. Flammable. Produces very toxic fumes when heated to decomposition. Exposure to small amounts over a long period may damage kidneys and liver and may cause dermatitis.
TLV (skin) 5 ppm (22 mg m^{-3}).
Flash point 95°C.
Autoignition temperature 559°C.

Incompatibility Oxidising agents.

Handling Wear gloves and eye protection.

Storage Flammables store.

Disposal Emulsify small quantities with water and detergent and wash to waste with water.

Spillage Apply water and detergent, brush to emulsify and proceed as for disposal. Consult Local Authority if spillage is large.

First Aid

Eyes Irrigate with water. Seek medical attention.

Lungs Remove patient from exposure, rest and keep warm. Seek medical advice.

Mouth Give water to drink. Seek medical attention.

Skin Wash with water and gently rub glycerol on affected area. Remove and wash contaminated clothing. Seek medical attention.

Local Conditions

$(CH_3)_2CHCH_2OH$ Liquid, BP 106°C.

Hazards Harmful vapour. Irritant to eyes and nose. Liquid harmful by skin absorption, causing internal injury. Flammable.
TLV 50 ppm (150 mg m^{-3}).
Flash point 28°C.
Autoignition temperature 427°C.

Incompatibility Oxidising agents.

Handling Wear gloves and eye protection. In fume cupboard preferably; if not available, in *well*-ventilated room away from all sources of ignition.

Storage Flammables store.

Disposal Unavoidable discharges, e.g. small quantities like washings from glassware, should be emulsified and washed to waste.

Spillage Turn off all sources of ignition. Wear gloves and face shield. Emulsify with detergent and water and run to waste. Alternatively absorb on sand, shovel into buckets for removal to open area for evaporation. Spillage area must be thoroughly ventilated.

First Aid

Eyes Irrigate with water. Seek medical advice.

Lungs Remove patient from exposure, rest and keep warm. Seek medical advice.

Mouth Wash out thoroughly with water. Seek medical advice.

Skin Wash thoroughly with water. Remove and wash contaminated clothing.

Local Conditions

Schedule 1 poison. A solution of mercury (II) nitrate in concentrated nitric acid.

Hazards The combination of the corrosive acid and the toxic mercury (II) nitrate combined with the fact that it is frequently boiled with the substance being tested makes its use extremely hazardous. See hazards associated with mercury compounds and nitric acid. Proteins containing tyrosine give a positive test with Millon's reagent. There are available less hazardous reagents which respond to other amino acids. These include Albustix, Cole's modification of Millon's reagent (uses mercury (II) sulphate in 2M sulphuric acid), and the Sakaguchi test.

Incompatibility Ethyne (acetylene), ethanedioic acid (oxalic acid), ethanol, hydrocarbons, sulphur, propanone (acetone).

Handling Wear gloves and eye protection. Often a satisfactory positive result can be obtained by merely heating the mixture of reagent and suspected protein by means of a water bath.

Storage Poisons cupboard.

Disposal Small amounts may be neutralised and washed to waste with a large volume of water. If amounts more than small consult Local Authority.

Spillage Sprinkle on sodium carbonate to neutralise the acid. Collect in a suitable container and consult Local Authority.

First Aid

Eyes Irrigate with water. Seek medical attention.

Lungs Remove patient from exposure, rest and keep warm. Seek medical advice.

Mouth Wash out with water. If swallowed give water to drink followed by milk of magnesia. Seek medical attention.

Skin Wash well with water. Remove contaminated clothing. Seek medical advice.

Local Conditions

Colourless to yellow liquid containing benzene, methylbenzene (toluene), dimethylbenzenes (xylenes), BP 150–216°C.

Hazards Because of its benzene content, mineral naphtha should not be available in any educational laboratory. For covering alkali metals use liquid paraffin. Poisonous by inhalation, swallowing or skin absorption. A recognised carcinogen (Sax). Flammable liquid. Explosion hazard slight.
Flash point 38°C.
Autoignition temperature 522°C.

Incompatibility Oxidising agents.

Handling With gloves and eye protection, in well-ventilated area preferably fume cupboard away from flames, hot plates, etc.

Storage Flammables store.

Disposal Small quantities such as washings from glassware, etc. should be emulsified and washed to waste.

Spillage Wear respirator, face shield and gloves. Switch off all heating sources. Mop up area of spillage with detergent and water. Mop and wash emulsion to waste with water or spread out on open ground for evaporation. Wash area of spillage with water and more detergent.

First Aid

Eyes Wash with water. Seek medical advice.

Lungs Remove patient from area, rest and keep warm. Seek medical advice.

Mouth Wash with water. Seek medical advice.

Skin Wash with soap and water. Seek medical advice.

Local Conditions

$C_{10}H_8$ White solid, MP 80°C.

Hazards Harmful by inhalation, ingestion and by skin contact. Fire hazard moderate when heated or on contact with flames. When used to show states of matter by heating in a test-tube, quantity should be small and heating gentle so that minimum of vapour escapes from test-tube. Safer alternatives for this experiment include hexadecan-1-ol, octadecan-1-ol, hexadecanoic acid and octadecanoic acid.
TLV 10 ppm (50 mg m^{-3}).
Flash point 120°C.
Autoignition temperature 526°C.

Incompatibility Strong oxidising agents.

Handling Avoid contact with skin; wear eye protection and gloves.

Storage General store or laboratory.

Disposal Wet with water and place in polythene bag in refuse bin.

Spillage As *Disposal*.

First Aid

Eyes Irrigate with water. Seek medical advice.

Lungs Remove patient to fresh air. Seek medical advice.

Mouth Wash with water. Seek medical advice.

Skin Wash with soap and water.

Local Conditions

$C_{10}H_7OH$ White solids, MPs 96 and 122.5°C.

Hazards Harmful by inhalation, swallowing or skin contact. Naphthalen-2-ol is slightly less toxic than naphthalen-1-ol. Absorption through skin may cause irritation of the kidneys and injury to cornea and lens of the eye.
Flash points of about 200°C. Fire hazard therefore slight.

Incompatibility Strong oxidising agents.

Handling Use gloves and eye protection. Handle in well-ventilated area.

Storage General store.

Disposal Small amounts: moisten and put in polythene bag in refuse bin.

Spillage Moisten and treat as in disposal.

First Aid

Eyes Irrigate with water. Seek medical advice.

Lungs Remove patient from area. Rest and keep warm. Seek medical advice.

Mouth Wash out with water. Seek medical attention.

Skin Wash with cold water and then with soap and water. Seek medical advice.

Local Conditions

Schedule 1 poison. This is a solution of mercury (II) iodide, potassium iodide and sodium hydroxide.

Hazards Hazardous components are the poison, mercury (II) iodide, and the corrosive sodium hydroxide.

Incompatibility Ethyne (acetylene), ethanedioic acid (oxalic acid), ethanol and hydrocarbons.

Handling Wear gloves and eye protection.

Storage Poisons cupboard.

Disposal Small amounts may be neutralised and washed to waste with a large volume of water. If amounts more than small consult Local Authority.

Spillage Wear gloves and eye protection. Neutralise with dilute sulphuric acid, mop up and if small wash to waste. If large consult Local Authority.

First Aid

Eyes Irrigate with water. Seek medical attention.

Lungs ——

Mouth Wash out with water. If swallowed give water and milk to drink. Seek medical attention.

Skin Wash well with water. Remove contaminated clothing. Seek medical advice.

Local Conditions

Green crystals.

Hazards The dust is harmful to eyes, skin and lungs. Solids and solutions are also harmful if swallowed. Recognised carcinogens (Sax). *TLV* of metal dust and soluble compounds is 0.1 mg m^{-3}.

Handling Avoid raising dust and contact of solids or solution with the skin. Wear gloves and eye protection.

Storage General store.

Disposal If soluble dissolve in water and wash to waste with water. If insoluble mix with sand, put in a polythene bag and into refuse bin.

Spillage As for disposal. Wear dust respirator.

First Aid

Eyes Irrigate with water. Seek medical advice.

Lungs Remove patient from exposure. Seek medical advice.

Mouth Wash with water. Seek medical attention.

Skin Wash with water.

Local Conditions

$C_9H_4O_3, H_2O$ Pale yellow crystals.

Hazards An irritant poison. Toxicity details unclear but biologically active and requires careful handling. This compound is generally obtained in spray form from an aerosol spray and there is great danger of accidental spraying of face, etc. Spray from aerosol atomiser is very flammable due to solvent (usually butan-1-ol).

Incompatibility Oxidising agents.

Handling Use only in a fume cupbard and never in the open laboratory. Gloves should be worn when using spray or handling the chemical in making up solution.

Storage General store.

Disposal Aerosol by slow release in a fume cupboard or in the open.

Spillage Wearing eye protection and gloves mop up with detergent and water, and wash to waste with large quantities of water.

First Aid

Eyes Irrigate thoroughly with water. Seek medical attention.

Lungs Remove patient from area, rest and keep warm. Seek medical advice.

Mouth Wash thoroughly with water. Seek medical advice.

Skin Wash with cold water, then soap and water.

Local Conditions

NO_3^-

Hazards	Large amounts taken by mouth may have serious or even fatal effects. The toxicity is also dependent on the cation. Fire hazard is moderate by spontaneous chemical action. Explosion hazard when exposed to heat or flame or shock (especially ammonium nitrate). Small repeated doses of nitrates lead to headaches, depression and mental impairment.
Incompatibility	When mixed with reducing substances, e.g. carbon, ammonium salts and organic powders, produce explosive mixtures.
Handling	In general no special precautions other than to avoid much friction.
Storage	General store or laboratory.
Disposal	This depends mainly on the metal present. Most nitrates such as those of K, Na, Ca, can be washed to waste. If amounts very large consult Local Authority.
Spillage	Small quantities can be washed to waste; otherwise moisten with ample water, mix with sand and shovel into normal waste bin in suitable container.

First Aid

Eyes	Irrigate with water. Seek medical attention.
Lungs	——
Mouth	Wash thoroughly with water. Seek medical advice.
Skin	Wash with water.

Local Conditions

HNO_3 Colourless or yellow fuming liquid. Concentrated nitric acid is about 70% by weight of pure nitric acid. Fuming nitric acid is 95–98% concentration and is much more corrosive than concentrated nitric acid. The latter should not be necessary in school chemistry. Considerable heat of dilution with water.

Hazards   Liquid and vapour irritant to eyes, lungs and skin.
TLV 2 ppm (5 mg m^{-3}).

Incompatibility   Dangerous if mixed with ethanoic acid (acetic acid), propanone (acetone), carbon, phosphorus, sulphur, chromic (VI) acid, ethanol, hydrogen sulphide, methanol, oxidisable organic materials such as paper, wood, readily nitrated materials.

Handling   Use gloves and eye protection. Dilution by careful addition of acid to water. The dilute acid should also be handled with great care.

Storage   With acids in general store, but not close to hydrobromic and hydriodic acids.

Disposal   Add very slowly to a large quantity of water, neutralise with sodium carbonate and wash to waste with running water.

Spillage   Wear gloves and face shield. Spread anhydrous sodium carbonate over the acid, mop up with plenty of water and wash to waste with plenty of water. Mop the area of spillage thoroughly with water.

First Aid

Eyes   Irrigate with water. Seek medical attention.

Lungs   Remove patient from exposure, rest and keep warm. Seek medical advice.

Mouth   Wash out with water. If swallowed give water to drink followed by milk of magnesia. Seek medical attention.

Skin   Wash thoroughly with water. Remove contaminated clothing and wash. Seek medical advice.

Local Conditions

NO_2^-

Hazards	Metal nitrites are very poisonous by ingestion and inhalation. Even small doses can prove fatal and prolonged exposure to very small doses is dangerous. They can explode by friction in contact with organic matter, when heated, or in contact with cyanides.
Incompatibility	Acids, cyanides, organic compounds.
Handling	Wear gloves and eye protection.
Storage	Poisons cupboard.
Disposal	Small amounts: wash to waste with plenty of water. For large amounts consult Local Authority.
Spillage	Mop up with plenty of water and wash to waste with a large volume of water. Wash the area of spillage well.

First Aid

Eyes	Irrigate with water. Seek medical advice.
Lungs	Remove patient from exposure, rest and keep warm. Seek medical advice.
Mouth	Wash with water. Seek medical attention.
Skin	Wash thoroughly with water.

Local Conditions

$C_6H_5NO_2$ Yellow liquid, BP 211°C. Odour of bitter almonds.

Hazards   Vapour is very harmful causing a burning sensation in the chest and difficulty in breathing. The liquid burns the eyes, poisons by skin absorption and is very poisonous when swallowed. Prolonged exposure via ingestion, inhalation and skin absorption is very dangerous. Flammable.
TLV (skin) 1 ppm (5 mg m^{-3}).
Flash point 88°C.
Autoignition temperature 522°C.

Incompatibility   Oxidising agents.

Handling   Wear gloves and eye protection. In well-ventilated area for small quantities but larger quantities in fume cupboard.

Storage   Flammables store.

Disposal   Unavoidable discharges, e.g. small quantities such as washings from glassware, etc. should be emulsified and washed to waste.

Spillage   Evacuate room and wear eye protection, a respirator and gloves. Apply water and detergent. Mop to emulsify and wash to waste with running water. Alternatively absorb on sand and transfer to open area for evaporation.

First Aid

Eyes   Irrigate with water. Seek medical attention.

Lungs   Remove patient from area, rest and keep warm. Seek medical advice.

Mouth   Wash mouth with water. Seek medical attention.

Skin   Wash with soap and water. Wash contaminated clothing. Seek medical advice.

Local Conditions

$C_6H_4(OH)(NO_2)$
2- or o- isomer, MP 45°C.
3- or m- isomer, MP 97°C.
4- or p- isomer, MP 113°C.
All are pale yellow solids slightly soluble in water. The 2- isomer is more volatile than the other two.

Hazards Very harmful both by acute dosage and by prolonged exposure to low concentrations. Dust and vapour irritate eyes and skin. Absorption by inhalation of dust, by skin, or by ingestion.

Incompatibility Strong oxidising agents, nitric acid.

Handling Wear eye protection and gloves. Handle in well-ventilated area or fume cupboard. Avoid raising dust.

Storage Poisons cupboard.

Disposal Wear gloves and eye protection. Small amounts such as washings from glassware may be run to waste.

Spillage Wear gloves and face shield. Shovel solid into bucket or large beaker and add a little sodium hydroxide solution to aid dissolution. Wash this solution to waste if quantity small. If large consult Local Authority about disposal.

First Aid

Eyes Irrigate with water. Seek medical attention.

Lungs Remove patient from exposure, rest and keep warm. Seek medical advice.

Mouth Wash out with water. If any has been swallowed give plenty of water to drink. Seek medical attention.

Skin Wash thoroughly with water. Seek medical attention immediately. Wash and air contaminated clothing.

Local Conditions

O Octane

$CH_3(CH_2)_6CH_3$ Colourless liquid, BP 125°C.

Hazards   Dangerous if inhaled. May act as asphyxiant. Highly flammable. Fire hazard similar to petrol.
TLV 300 ppm (1450 mg m⁻³).
Flash point 13°C.
Autoignition temperature 260°C.

Incompatibility   Strong oxidising agents.

Handling   Wear eye protection and keep away from flames, hot plates, etc. Use in well-ventilated conditions.

Storage   Flammables store. Up to 250 cm³ in laboratory.

Disposal   Unavoidable discharges, e.g. small quantities such as washings from glassware, etc. should be emulsified and washed to waste.

Spillage   Turn off flames and heaters. Wearing gloves and eye protection, add detergent and water, emulsify with mop and transfer to bucket. Spread out in open area to evaporate.

First Aid

Eyes   Wash with water. Seek medical advice.

Lungs   Remove patient to fresh air. Seek medical advice.

Mouth   Wash out mouth with water. Seek medical advice.

Skin   Wash with soap and cold water.

Local Conditions

C_8H_{16} Colourless liquid, BP 121°C.

Hazards Flammability main danger.
Flash point 21°C.

Incompatibility Oxidising agents.

Handling Well away from sources of ignition with good ventilation. Eye protection advisable.

Storage Flammables store. Up to 250 cm³ in laboratory.

Disposal Unavoidable discharges, e.g. small quantities such as washings from glassware, etc. should be emulsified and washed to waste.

Spillage Turn off all sources of ignition. Wearing gloves and eye protection, add water and detergent. Emulsify by mopping and wash to waste with running water if quantity is small. If quantity is large, absorb on sand, brush up, collect in a bucket and transfer to open area for evaporation.

First Aid

Eyes Wash with water. Seek medical advice.

Lungs Remove patient from area, rest and keep warm.

Mouth Wash with water. Seek medical advice.

Skin Wash with soap and water.

Local Conditions

O Oleum
Fuming sulphuric acid

Colourless to yellow viscous fuming liquid.

Hazards This is up to 80% solution of sulphur (VI) oxide in sulphuric acid. Extremely dangerous liquid and fumes. Causes burns to eyes and skin, and fumes are very harmful to all parts of respiratory system. Extremely irritant if swallowed. This should not normally be used in a school chemistry laboratory. Much more dangerous than concentrated sulphuric acid.

Incompatibility Reacts very violently with water or substances containing water, and with reducing agents.

Handling Wear face shield, gloves and PVC apron. Use fume cupboard.

Storage With acids in general store.

Disposal Do this in fume cupboard, adding very slowly to a large quantity of water, neutralise with sodium carbonate and wash to waste with plenty of running water.

Spillage Wear face shield, gloves and PVC apron. Spread anhydrous sodium carbonate over the spillage and mop up with large quantities of water. Wash to waste with running water.

First Aid

Eyes Irrigate with water. Seek medical attention.

Lungs Remove patient from area, rest and keep warm. Seek medical advice.

Mouth Wash with water. If swallowed give water and milk of magnesia to drink. Seek medical attention.

Skin Wash thoroughly with water and apply magnesia/glycerol paste. Contaminated clothing would probably be burned and not be suitable for re-use. Seek medical advice.

Local Conditions

OsO_4 Colourless to yellow crystals, MP 40°C.

Hazards   So poisonous and expensive that its use is not recommended. Vapour irritates the respiratory system. Vapour, solid and solution burn the eyes severely. Acid and solution burn the skin, and continued exposure to vapour causes dermatitis and distortion of vision. Very poisonous if swallowed.
TLV 0.0002 ppm (0.002 mg m^{-3}).

Incompatibility   When heated emits toxic fumes.

Handling   Supplied in ampoules which should be opened in fume cupboard. Follow instructions, wear gloves and eye protection. A 1% solution is often supplied and this can be used in open laboratory but wearing gloves and eye protection.

Storage   Poisons cupboard.

Disposal   Wash to waste with running water.

Spillage   Wear gloves and face shield. Mop up with plenty of water and wash to waste with running water.

First Aid

Eyes   Irrigate with water. Seek medical attention.

Lungs   Remove patient to fresh air. Rest and keep warm. Seek medical advice.

Mouth   Wash with water. If swallowed drink water followed by milk of magnesia. Seek medical attention.

Skin   Wash with soap and water. Seek medical advice.

Local Conditions

O Oxygen mixture
Potassium chlorate(V) and manganese(IV) oxide

$KClO_3 + MnO_2$ This mixture should ideally neither be kept nor made up in educational laboratories.

Hazards May explode violently on heating if contaminated by reducing agents including carbon, organic compounds or even by dust. Toxic owing to the presence of potassium chlorate (V).

Incompatibility Any reducing agent, e.g. carbon, metals, sulphuric acid.

Handling Use gloves, face shield and explosion screen. Use only small amounts at a time. Avoid risk of contamination by using clean pyrex glassware, spatulae, etc.

Storage General store with oxidising agents.

Disposal Wash small amounts to waste with water.

Spillage Wear face shield and gloves. Dampen slightly and carefully sweep up with a soft brush. Add to plastic bucket containing water. Wash to waste with plenty of water. Site of spillage should be well washed with water.

First Aid

Eyes Irrigate with water. Seek medical attention.

Lungs ——

Mouth Rinse thoroughly with water. Seek medical attention.

Skin Wash with water and then with soap and water. Seek medical advice especially if skin has been broken.

Local Conditions

$CH_3(CH_2)_3CH_3$ Colourless liquid, BP 36°C.

Hazards Vapour is narcotic if concentration is high. Harmful if liquid swallowed or gets into eyes. Highly flammable.
TLV 600 ppm (1800 mg m^{-3}).
Flash point −48°C.
Autoignition temperature 309°C.

Incompatibility Oxidising agents. Explosion hazard if exposed to heat or flame.

Handling Wear gloves and eye protection. In well-ventilated area, well away from flames and hot plates, etc.

Storage Flammables store. Up to 250 cm^3 in laboratory.

Disposal Unavoidable discharges, e.g. small quantities such as washings from glassware, etc. should be emulsified and washed to waste.

Spillage Turn off all sources of ignition, wear gloves and face shield. Add water and detergent, mop to emulsify and wash to waste with running water. Alternatively the liquid can be absorbed on sand, shovelled into a bucket and spread out in an open area for evaporation.

First Aid

Eyes Irrigate with water. Seek medical advice.

Lungs Remove patient to fresh air, rest and keep warm.

Mouth Wash with water. Seek medical advice.

Skin Wash with soap and water.

Local Conditions

Pentan-1-ol and pentan-2-ol
n- and sec-Amyl alcohol

$C_5H_{11}OH$ Colourless liquids, BPs 138°C and 119°C.

Hazards   More toxic than ethanol. Harmful vapour to eyes and respiratory system. Liquid if swallowed is very harmful causing giddiness, etc. Prolonged exposure to low concentrations dangerous. Flammable.
Flash points 33°C and 40°C.
Autoignition temperature 300°C.

Incompatibility   Reacts with strong oxidising agents.

Handling   In well-ventilated area, well away from sources of ignition. Gloves and eye protection advisable.

Storage   Flammables store. Up to 250 cm³ in laboratory.

Disposal   Unavoidable discharges, e.g. small quantities such as washings from glassware, etc. should be emulsified and washed to waste.

Spillage   Turn off all sources of ignition. Wear gloves and face shield. Apply water and detergent. Mop to emulsion and wash to waste with running water. Alternatively the liquid can be absorbed in sand, shovelled into a bucket and spread out in an open area for evaporation.

First Aid

Eyes   Irrigate with water. Seek medical attention.

Lungs   Remove patient from area, rest and keep warm. Seek medical advice.

Mouth   Wash mouth with water and give water to drink. If appreciable amount swallowed seek medical advice.

Skin   Wash with soap and water.

Local Conditions

$C_2H_5COC_2H_5$ Liquid, BP 101°C.

Hazards Toxicity slight. Reacts with some oxidising agents. Highly flammable.
Flash point 13°C.
Autoignition temperature 452°C.

Incompatibility Oxidising agents.

Handling In well-ventilated area, away from flames, hot gauze, hot plates, etc.

Storage Flammables store.

Disposal Small quantities, mix with water and detergent to emulsify and wash to waste with water.

Spillage Turn off all sources of ignition in the room. Evacuate room if spillage is large. Mop up with plenty of water and detergent and wash to waste. Ventilate the room thoroughly.

Fist Aid

Eyes Irrigate with water. Seek medical advice.

Lungs Remove patient from area of contamination, rest and keep warm.

Mouth Wash with water. Drink water. Seek medical advice.

Skin Wash with soap and water.

Local Conditions

P Pentyl ethanoate
Amyl acetate

$CH_3COOC_5H_{11}$ Liquid, BP 148°C. Pear drop smell.

Hazards Vapour is irritant to eyes, lungs, skin. Liquid is moderately toxic. Highly flammable.
TLV 100 ppm (525 mg m^{-3}).
Flash point 25°C.
Autoignition temperature 379°C.

Incompatibility Dangerous when exposed to heat or flame. Reacts vigorously with reducing agents.

Handling Wear eye protection and use in well-ventilated area and well away from open flames, hot plates, etc.

Storage Flammables store. Up to 250 cm^3 in laboratory.

Disposal Unavoidable discharges, e.g. small quantities such as washings from glassware, etc. should be emulsified and washed to waste.

Spillage Turn off all sources of ignition. Wear gloves, face shield and respirator. Apply water and detergent to disperse the liquid. If quantity small, mop up and wash to waste with running water. For larger quantities spread over open ground.

First Aid

Eyes Irrigate with water. Seek medical advice.

Lungs Remove patient from area. Rest and keep warm.

Mouth Wash with water. Seek medical advice.

Skin Wash with water.

Local Conditions

Hazards	Metal present determines toxicity, e.g. barium peroxide is a Schedule 1 poison. Cause injury on contact with skin and mucous membranes. Strong oxidising agents.
Incompatibility	Fire hazard varies from moderate to very dangerous when in contact with reducing agents and contaminants. Explosion hazard with heat, shock or catalyst. React with metal powders, non-metals, organic substances, water, steam and acids, and may emit toxic fumes.
Handling	Wear gloves and eye protection. Keep away from moisture and out of contact with organic materials. Use the smallest quantity necessary for the experiment.
Storage	With oxidising agents. Except in the case of hydrogen peroxide solution, the stopper should be tight to avoid moisture reaching the chemical.
Disposal	If quantity is small, add a little at a time to a large quantity of water and wash to waste.
Spillage	*For other than scheduled poisons* Mix with dry sand, shovel into suitable dry container, remove to an open area and scatter sparingly on the gound. Alternatively treat as in disposal. *For scheduled poisons* Mix with dry sand, shovel into suitable dry container and consult Local Authority about disposal.

First Aid

Eyes	Irrigate with water. Seek medical attention.
Lungs	Seek medical advice.
Mouth	Wash out with water. Seek medical attention.
Skin	Brush off powder as well as possible and wash with cold water.

Local Conditions

Petroleum spirit
Petroleum ether

Colourless liquid. Fractions with BPs in variety of ranges from 30–40°C up to 120–160°C.

Hazards High concentrations of vapour cause intoxication. Liquids irritate skin and eyes. Very harmful if swallowed. Highly flammable.
TLV 500 ppm (2000 mg m^{-3}).
Flash point −17°C.
Autoignition temperature 290°C for the lowest boiling fraction.

Incompatibility Chlorine, oxidising agents.

Handling In well-ventilated area, well away from sources of ignition. Wear eye protection and gloves since liquid dissolves fat in the skin.

Storage Flammables store. Up to 250 cm^3 in laboratory.

Disposal Unavoidable discharges, e.g. small quantities such as washings from glassware, etc. should be emulsified and washed to waste.

Spillage Turn off all sources of ignition. Wear gloves and face shield. Mop up with water and detergent to emulsify. Wash to waste with running water. Alternatively absorb in sand, shovel into bucket and spread on open ground for evaporation. Allow any liquid remaining at spillage site to evaporate.

First Aid

Eyes Irrigate with water. Seek medical advice.

Lungs Remove patient from area, rest and keep warm.

Mouth Wash with water. Seek medical advice.

Skin Wash with soap and water immediately.

Local Conditions

C_6H_5OH Colourless to pink crystals, MP 43°C.

Hazards Vapour is harmful to the eyes, lungs and skin. Solid or solution is very poisonous if swallowed. Solid and solution are very corrosive, causing whitening of the skin. Poisonous by skin absorption. Prolonged exposure to low concentrations of mist or vapour very dangerous. A cocarcinogen.
TLV (skin) 5 ppm (19 mg m^{-3}).

Incompatibility Oxidising agents.

Handling Gloves and eye protection must be worn when handling solid or solution.

Storage Poisons cupboard.

Disposal Wash small amounts such as washings from glassware, etc. to waste with running water.

Spillage Wear gloves and face shield. If spillage is small, shovel solid into a bucket of water and when dissolved wash to waste with running water. If spillage is large consult Local Authority about disposal.

First Aid

Eyes Irrigate with water. Seek medical attention.

Lungs Remove patient from exposure, rest and keep warm. Seek medical attention.

Mouth Wash out with water and if any has been swallowed, drink plenty of water. Seek medical attention.

Skin Wash thoroughly with soap and water. Rub with propane-1,2,3-triol (glycerol) and seek medical attention immediately. Wash contaminated clothing.

Local Conditions

P Phenolphthalein

$C_{20}H_{14}O_4$ White powder, MP 261°C.

Hazards Ingestion is harmful causing stomach upset, diarrhoea. It is a solid but generally used in solution in ethanol as acid/base indicator.
The solution is flammable.

Incompatibility Strong oxidising agents.

Handling No special precautions required.

Storage General store.

Disposal Emulsify by adding water and detergent, and wash to waste with water.

Spillage Add water and detergent and mop up. Wash to waste.

First Aid

Eyes Irrigate with water. Seek medical advice.

Lungs ——

Mouth Wash out. Seek medical advice.

Skin Wash with water.

Local Conditions

$C_6H_5NH_2$ Colourless to brown liquid, BP 185°C.

Hazards Very poisonous if inhaled, swallowed or absorbed through skin. Liquid is harmful to the eyes. Flammable. Long term exposure to low concentrations affects nervous system and blood.
TLV (skin) 5 ppm (19 mg m^{-3}).
Flash point 70°C.
Autoignition temperature 770°C.

Incompatibility Oxidising agents especially hydrogen peroxide; nitric acid.

Handling Wear gloves and eye protection. In well-ventilated area, preferably a fume cupboard, and well away from flames or hot plates, etc. Distillations and experiments in fume cupboard.

Storage Flammables store. Should not be stored in laboratory.

Disposal Small quantities such as washings from glassware should be neutralised with dilute hydrochloric acid, emulsified with water containing a dispersing agent, and put down the drain with plenty of running water.

Spillage Wearing respirator, gloves and face shield, absorb in sand, shovel into closable container and consult Local Authority for disposal. Mop area of spillage with water and detergent, neutralise washings and put down the drain with running water.

First Aid

Eyes Irrigate with slow flow of water. Seek medical advice.

Lungs Rest and keep warm. Seek medical advice.

Mouth Wash out mouth with water. Seek medical advice.

Skin Remove contaminated clothing. Wash the skin at once with water then soap and water. Seek medical advice.

Local Conditions

P Phenylammonium chloride
Aniline hydrochloride

$C_6H_5NH_3Cl$ White to reddish crystalline solid, MP 200°C.

Hazards Poisonous by swallowing, skin absorption and by inhalation. Acute effects are headache, drowsiness, cyanosis, mental confusion and, in severe cases, convulsions. Exposure over a long period affects the nervous system and the blood. Dangerous to the eyes, partly because of acidity, causes intense irritation. *Flash point* 193°C.

Incompatibility Oxidising agents especially hydrogen peroxide, nitric acid.

Handling Wear gloves and eye protection. Use in fume cupboard.

Storage Flammables store. Should not be stored in laboratory.

Disposal Wash small amounts to waste with water.

Spillage Shovel into bucket and wash to waste with water. Ensure that dilution is adequate.

First Aid

Eyes Irrigate with water. Seek medical advice.

Lungs ——

Mouth If swallowed wash out mouth thoroughly with water. Seek medical advice.

Skin Drench with water and wash with soap and water; remove and wash contaminated clothing before re-use.

Local Conditions

$C_6H_5CHCH_2$ Colourless liquid, BP 146°C.

Hazards Vapour has disagreeable odour and is harmful to respiratory system and eyes. Liquid is poisonous if swallowed and is harmful to the skin. Highly flammable.
TLV 100 ppm (420 mg m^{-3}).
Flash point 31°C.
Autoignition temperature 490°C.

Incompatibility Heating produces acrid fumes. Reacts vigorously with oxidising agents.

Handling Wear gloves and eye protection. In fume cupboard, away from sources of ignition.

Storage Flammables store. Never in laboratory.

Disposal Wear gloves and eye protection. Unavoidable discharges, e.g. small quantities such as washings from glassware, etc. should be emulsified and washed to waste.

Spillage Wear gloves and face shield. Turn off all sources of ignition. Add water and detergent, mop to emulsify and put down drain with plenty of running water. Alternatively sand can be used to absorb the liquid. Shovel into plastic bucket and spread out in open area for evaporation.

First Aid

Eyes Irrigate with water. Seek medical attention.

Lungs Remove patient from area, rest and keep warm. Seek medical advice.

Mouth Wash with water. Seek medical attention.

Skin Wash with soap and water. Seek medical advice.

Local Conditions

$C_6H_5NHNH_2$ Yellow to red-brown liquid or solid, MP 19°C.
$C_6H_5NHNH_3Cl$ Solid, MP 245°C.

Hazards The solid is harmful in dust or vapour form to skin and eyes. Very harmful by skin absorption and if swallowed or dust inhaled. Dangerous when heated to decomposition. Flammable.
TLV (skin) vapour or dust 5 ppm (22 mg m^{-3}).
Flash point 89°C.

Incompatibility Oxidising agents.

Handling In well-ventilated laboratory, wearing gloves and eye protection. Best handled in solution by pupils.

Storage General store.

Disposal Add the solid to dilute hydrochloric acid and leave for 24 hours. Finally wash to waste with water.

Spillage Wear gloves and face shield. Mix the solid with sand and shovel into plastic bucket. Add excess of dilute hydrochloric acid and leave for 24 hours. Wash to waste with running water. The sand can be put in normal waste bin.

First Aid

Eyes Irrigate with water. Seek medical attention.

Lungs Remove patient to fresh air, rest and keep warm. Seek medical attention.

Mouth Wash with water. Seek medical attention.

Skin Wash with soap and water. Seek medical advice.

Local Conditions

$C_6H_5CH_2OH$ Liquid, BP 206°C.

Hazards Harmful if inhaled or swallowed. Vapour or liquid is allergen, causing dermatitis. It can cause headaches, nausea, vomiting and diarrhoea. Fire hazard is slight when exposed to heat or flame.
Flash point 100.5°C.
Autoignition temperature 434°C.

Incompatibility Oxidising agents.

Handling In well-ventilated area. Wear gloves and eye protection.

Storage General store. Up to 250 cm³ in laboratory.

Disposal Small quantities such as washings from glassware, etc. should be emulsified and washed to waste.

Spillage Mop area of spillage with water and detergent and wash to waste with running water.

First Aid

Eyes Irrigate with water. Seek medical advice.

Lungs Rest and keep warm. Seek medical advice.

Mouth Wash out mouth with water. Seek medical advice.

Skin Wash the skin with water at once. Seek medical advice.

Local Conditions

P Phosphoric (V) acid
Orthophosphoric acid

H_3PO_4 Colourless liquid (about 90%) or moist crystals (100%).

Hazards Burns the skin and eyes. If swallowed causes serious internal injury. Dangerous when heated to decomposition, giving oxides of phosphorus.
TLV 1 mg m^{-3}.

Incompatibility Water; though not a strong acid it is available in high concentration (90% or 16M) and the heat of dilution is high.

Handling Wear gloves and eye protection.

Storage With acids in general store.

Disposal Add to solution of sodium carbonate to neutralise and wash to waste with running water.

Spillage Wearing face shield and gloves spread anhydrous sodium carbonate over the spillage and mop up with plenty of water. Wash to waste with running water.

First Aid

Eyes Irrigate with water. Seek medical attention.

Lungs ——

Mouth Wash with water. Drink water and then milk of magnesia. Seek medical attention.

Skin Wash with water and apply magnesia/glycerol paste. Seek medical advice.

Local Conditions

Red powder subliming at 416°C.

Hazards Flammable solid. On burning it yields highly toxic fumes. Harmful if in contact with skin or eyes.
Autoignition temperature 260°C.

Incompatiblity Oxidising agents, e.g. halogens, manganates (VII) (permanganates), reactive metals and alkalis.

Handling Wear gloves and eye protection.

Storage General store, away from alkalis and from both oxidising and reducing agents.

Disposal Small amounts by burning off in a crucible in a fume cupboard.

Spillage Wear gloves and face shield, cover with wet sand and shovel into bucket. Remove to safe open area and burn away after the phosphorus has dried.

First Aid

Eyes Irrigate with water. Seek medical attention.

Lungs If vapour of burning phosphorus inhaled, remove patient from exposure, rest and keep warm. Seek medical advice.

Mouth Wash out thoroughly with water. Seek medical advice.

Skin Wash thoroughly with water. Seek medical advice.

Local Conditions

Pale yellow translucent solid, MP 440°C.

Hazards Poisonous if swallowed or if vapour or smoke from burning phosphorus is inhaled. Long term absorption of small amounts by inhalation and by mouth is very dangerous. Extremely corrosive; should never be allowed to contact the skin. Highly flammable.
TLV 0.1 mg m^{-3}.
Flash point – spontaneously in air.
Autoignition temperature 30°C.

Incompatibility Oxidising agents, e.g. halogens (reaction of bromine and phosphorus is explosive), manganates (VII) (permanganates), alkalis and some metals.

Handling Wear gloves and eye protection. Use tongs. Never expose large pieces to the air. Cutting must be done under cold water. Keep under cold water till required.

Storage In locked cupboard in general store, away from alkalis and from oxidising and reducing agents. Must be stored under water.

Disposal Small amounts by burning off in crucible in a fume cupboard.

Spillage Wear gloves and face shield. Cover with wet sand, shovel phosphorus and sand into a bucket and cover with water. Remove to an open area for drying out and burning. Wash area of spillage with copper (II) sulphate and water.

First Aid

Eyes Irrigate with water. Seek medical attention.

Lungs Remove patient from exposure, rest and keep warm. Seek medical attention.

Mouth Wash out mouth with water. Seek medical attention.

Skin Wash with plenty of water. Swab with 3% solution of copper (II) sulphate to form a black copper salt which can be seen and removed. Seek medical advice.

Local Conditions

P_2O_5 White powder.

Hazards	Fine powder which is very irritant to the respiratory system and burns the eyes and skin. Very severe irritation if swallowed.
Incompatibility	Reacts violently with water or steam, reducing agents, especially organic compounds, alkali metals and alkaline earth metals.
Handling	Wear gloves and eye protection. Handle carefully to avoid dust rising.
Storage	With oxidising agents.
Disposal	Wear gloves and face shield. In the fume cupboard, sprinkle into water, neutralise with sodium carbonate and wash to waste with running water.
Spillage	Wear gloves and face shield. Mix with dry sand and shovel into plastic bucket. Remove to open area and add slowly to a large volume of water. Neutralise with sodium carbonate and put down the drain with running water.

First Aid

Eyes	Irrigate with water. Seek medical attention.
Lungs	Remove patient to fresh air, rest and keep warm. Seek medical advice.
Mouth	Wash with water. Drink water followed by milk of magnesia. Seek medical attention.
Skin	Wash with water, apply magnesia/glycerol mixture.

Local Conditions

PCl_5 White to pale yellow solid.

Hazards   Vapour and dust very harmful to respiratory system, eyes, skin and if swallowed. When exposed to air produces corrosive, toxic fumes of hydrogen chloride and oxoacids of phosphorus. Prolonged exposure to such fumes is harmful.
TLV 1 mg m^{-3}.

Incompatibility   Reacts violently with water, lower alcohols and acids, producing heat and toxic fumes of hydrogen chloride. Can react with some organic materials generating enough heat to ignite mixtures.

Handling   In fume cupboard or well-ventilated laboratory, wearing gloves and eye protection. Keep away from water and solutions containing water.

Storage   With hydrolysable halides, well-ventilated, and tightly stoppered.

Disposal   Wearing gloves, respirator and face shield, add a little at a time to a large volume of water. Neutralise with sodium carbonate in fume cupboard. Wash to waste with running water.

Spillage   Evacuate room and wear gloves, respirator and face shield. Mix with dry sand and shovel into plastic bucket. Remove to an open area and add a little at a time to a large volume of water. Neutralise with sodium carbonate. Wash solution to waste with running water.

First Aid

Eyes   Irrigate with water. Seek medical attention.

Lungs   Remove patient to fresh air, rest and keep warm. Seek medical advice.

Mouth   Wash with water. Drink water followed by milk of magnesia. Seek medical attention.

Skin   Wash with water. Treat with magnesia/glycerol paste.

Local Conditions

PCl_3 Colourless, fuming liquid.

Hazards   Vapour from the liquid reacts with moisture in air to form corrosive, toxic fumes including hydrogen chloride. Vapour and the toxic fumes produced are very irritating to eyes, respiratory system and skin. If swallowed causes severe internal damage. *TLV* 0.5 ppm (3 mg m^{-3}).

Incompatibility   Reacts vigorously with water, steam and acids to produce toxic fumes, and with strong oxidising agents.

Handling   In fume cupboard or well-ventilated laboratory, wearing gloves and eye protection. Keep away from water, steam, acids and alkalis.

Storage   With hydrolysable halides, well-ventilated, and tightly stoppered.

Disposal   In fume cupboard, add a little at a time to a large volume of water. Neutralise with sodium carbonate. Wash to waste with running water.

Spillage   Evacuate room and wear face shield, respirator and gloves. Mix with dry sand and shovel into plastic bucket. Remove to an open area and add a little at a time to a large volume of water. Neutralise with sodium carbonate. Wash solution to waste with running water.

First Aid

Eyes   Irrigate with water. Seek medical attention.

Lungs   Remove patient from area, rest and keep warm. Seek medical advice.

Mouth   Wash with water. Drink water followed by milk of magnesia. Seek medical attention.

Skin   Wash with water, and apply magnesia/glycerol paste.

Local Conditions

P Poly(methanal)
Paraformaldehyde

$(CH_2O)_x$ White crystalline powder, MP 64°C.

Hazards Harmful to eyes, to skin and to respiratory system, and if swallowed. Vaporises to methanal gas when heated. Flammable.
Flash point 70°C.
Autoignition temperature 340°C.

Incompatibility Strong oxidising agents. Reacts with hydrogen chloride to form the carcinogen chloromethoxychloromethane (bischloromethyl ether).

Handling Wear gloves and eye protection. In open laboratory if care is taken to avoid dust rising. Keep away from hydrogen chloride fumes.

Storage Flammables store.

Disposal Dissolve in water and wash to waste with water.

Spillage Wash to waste with plenty of running water.

First Aid

Eyes Irrigate with water. Seek medical attention.

Lungs Remove patient to fresh air, rest and keep warm. Seek medical advice.

Mouth Wash with water. Seek medical attention.

Skin Wash with water.

Local Conditions

Soft silvery metal.

Hazards Causes severe blisters on skin. Ignites spontaneously on exposure to moist air. Metal which has oxidised on storage should be disposed of since it my explode violently when cut. Highly flammable.

Incompatibility Water, carbon dioxide, oxidising agents, halogenated hydrocarbons, acids.

Handling Never touch the metal; use gloves, eye protection and tongs. Do not expose dry metal to air except for a few seconds.

Storage With reducing agents. Store in liquid paraffin.

Disposal Add to excess dry propan-2-ol and when reaction complete wash to waste with large excess of water.

Spillage Clear laboratory. Wear gloves and face shield. Cover with *dry* sand, shovel into dry container and transport to safe open area. Add in small quantities to excess dry propan-2-ol and wash to waste with excess running water.

First Aid

Eyes Flood with plenty of water. Seek medical attention.

Lungs ——

Mouth Wash with water. Seek medical attention.

Skin Remove any adhering metal and drench skin with water. Seek medical advice.

Local Conditions

P Potassium hexacyanoferrate (II)
Potassium ferrocyanide

$K_4Fe(CN)_6, 3H_2O$ Yellow crystals.

Hazards Dangerous if heated or if in contact with concentrated acids since it emits poisonous fumes of hydrogen cyanide. Moderately toxic itself.

Incompatibility Concentrated acids, dilute mineral acids.

Handling Use eye protection.

Storage General store, but away from acids.

Disposal Wash to waste with water.

Spillage As for disposal.

First Aid

Eyes Irrigate with water. Seek medical advice.

Lungs ——

Mouth Wash with water. Seek medical advice.

Skin Wash with water.

Local Conditions

$K_3Fe(CN)_6, 3H_2O$ Red crystals.

Hazards	If heated or in contact with concentrated acids it emits poisonous fumes of hydrogen cyanide. Toxic itself.
Incompatibility	Concentrated and dilute mineral acids, ammonia.
Handling	Use eye protection.
Storage	General store, away from acids and ammoniacal solutions.
Disposal	Wash to waste with water.
Spillage	As for disposal.

First Aid

Eyes	Irrigate with water. Seek medical advice.
Lungs	——
Mouth	Wash with water. Seek medical advice.
Skin	Wash with water.

Local Conditions

P Potassium hydroxide
Caustic potash

KOH   Colourless solid.

Hazards Skin contact causes severe blisters. Strongly corrosive solid and solution, very harmful if swallowed. Extremely dangerous to eyes.
(C) *TLV* 2 mg m^{-3}.

Incompatibility Acids, aluminium, zinc. With water produces considerable heat.

Handling Wear gloves and eye protection.

Storage General store.

Disposal Wear gloves and face shield. Dissolve in large volume of water, neutralise and wash to waste with running water.

Spillage Wear gloves and face shield. Add sand and shovel into dry plastic bucket for transport to safe open area and treat as in disposal.

First Aid

Eyes Irrigate with water. Seek medical attention.

Lungs ——

Mouth Wash thoroughly with water. If swallowed drink plenty of water. Seek medical advice.

Skin Drench with plenty of water. Wash contaminated clothing.

Local Conditions

$KMnO_4$ Dark purple crystals.

Hazards Strong irritant on skin. Dust harmful to lungs. Can explode on sudden heating and is common cause of eye accidents.

Incompatibility Being a powerful oxidising agent, it can cause ignition of organic matter especially propane-1,2,3-triol (glycerol), and of reducing agents including metals, sulphur, phosphorus, etc. With concentrated sulphuric acid it forms an explosive mixture. Hydrogen peroxide reacts violently.

Handling Wear gloves and eye protection. For class experiments demonstrating its thermal decomposition, heat small amounts of crystals in small, hard, glass test-tubes, preferably with a loose plug of ceramic wool (not Rocksil) to filter off solid particles, *not* in open on crucible lids, etc.

Storage General store with oxidising agents.

Disposal Easily soluble in water. Mop up with water and wash to waste with water. Wash area of spillage well.

Spillage As for disposal.

First Aid

Eyes Irrigate with water. Seek medical attention.

Lungs Remove patient from exposure, rest and keep warm.

Mouth Wash thoroughly with water. Seek medical advice.

Skin Wash thoroughly with water. Remove and wash contaminated clothing.

Local Conditions

P Potassium peroxodisulphate (VI)
Potassium persulphate

$K_2S_2O_8$ White solid.

Hazards May cause skin irritation and dermatitis. Solid and solution readily liberate oxygen on heating. Solid heated to decomposition gives oxides of sulphur. Harmful to lungs if swallowed.

Incompatibility Reducing agents, organic compounds.

Handling Wear gloves and eye protection.

Storage General store with oxidising agents.

Disposal Wash to waste with water, after neutralising with acidified iron (II) sulphate.

Spillage Moisten with water, shovel into bucket and treat as for disposal. Wash spillage area well.

First Aid

Eyes Irrigate with water. Seek medical attention.

Lungs Seek medical advice.

Mouth Wash with water. Seek medical advice.

Skin Wash with water.

Local Conditions

KSCN   Colourless deliquescent crystals.

Hazards   If heated strongly emits toxic fumes of cyanides. Decomposes at 500°C. Harmful if swallowed.

Incompatibility   Concentrated sulphuric acid, nitric acid.

Handling   Normal.

Storage   General store, away from acids.

Disposal   Wash to waste with water.

Spillage   Shovel into bucket and treat as for disposal.

First Aid

Eyes   Irrigate with water. Seek medical advice.

Lungs ——

Mouth   Wash with water. Seek medical advice.

Skin   Wash with water.

Local Conditions

CH_3CH_2CHO Colourless liquid, BP 49°C.

Hazards Liquid harmful to the skin. Vapour is very harmful to the eyes and lungs. Lachrymatory. Highly flammable, volatile liquid. *Flash point* −7°C.

Incompatibility Oxidising agents, alkalis.

Handling Wear gloves and eye protection. In a well-ventilated area, preferably a fume cupboard. Turn off all sources of ignition.

Storage Flammables store. Not in the laboratory.

Disposal Mix with water containing detergent and wash to waste with water.

Spillage Turn off all sources of ignition. Evacuate laboratory. Wear respirator, face shield and gloves. Add detergent and work to emulsion with brush and water. Mop up and wash to waste with running water. Alternatively absorb on sand, remove outdoors and allow to evaporate.

First Aid

Eyes Irrigate with water. Seek medical attention.

Lungs Remove patient from exposure, rest and keep warm. Seek medical advice.

Mouth Wash thoroughly with water. Seek medical advice.

Skin Drench skin with plenty of water. Wash contaminated clothing.

Local Conditions

CH_3CH_2COOH Colourless liquid, BP 141°C.

Hazards Liquid burns the skin and eyes. Vapour affects the lungs.
Flammable liquid.
Flash point 54°C.
Autoignition temperature 513°C.

Incompatibility Oxidising agents, alkalis, phosphorus pentachloride.

Handling In fume cupboard, away from any source of ignition. Wear gloves
and eye protection.

Storage Flammables store. Never in the laboratory.

Disposal Neutralise with sodium carbonate and wash down drain with
plenty of water.

Spillage Shut off all sources of ignition. Wear face shield, gloves and
respirator. Mop up and wash to waste with water. Ventilate
spillage area well.

First Aid

Eyes Irrigate with water. Seek medical attention.

Lungs Remove patient from exposure, rest and keep warm. Seek medical advice.

Mouth Wash thoroughly. If swallowed give plenty of water followed by
milk of magnesia. Seek medical advice.

Skin Drench with water. Wash contaminated clothing.

Local Conditions

Propan-1-ol and propan-2-ol
n- and iso-Propyl alcohol

$CH_3CH_2CH_2OH$ Colourless liquid, BP 97°C.
$CH_3CH(OH)CH_3$ Colourless liquid, BP 80°C.

Hazards Liquid and vapour of both are harmful to eyes and respiratory tract. Propan-2-ol is less toxic but is more volatile. Both are highly flammable.
TLV (skin) 200 ppm (500 mg m^{-3}) for propan-1-ol, propan-2-ol not available.
Flash point 15°C for propan-1-ol and 11°C for propan-2-ol.
Autoignition temperature 371°C for propan-1-ol and 405°C for propan-2-ol.

Incompatibility Strong oxidising agents, alkali metals, phosphorus pentachloride.

Handling In well-ventilated area with all sources of ignition turned off. Wear eye protection and gloves.

Storage Flammables store. Up to 250 cm³ in laboratory.

Disposal Wash to waste with plenty of water.

Spillage Turn off all possible sources of ignition. Wear gloves and face shield. Mop up and wash to waste with water. Ventilate spillage area well to disperse any vapour.

First Aid

Eyes Irrigate with water. Seek medical attention.

Lungs Remove patient from exposure.

Mouth Wash with plenty of water. Seek medical advice.

Skin Wash with water. Seek medical advice.

Local Conditions

CH_3COCH_3 Liquid, BP 56°C.

Hazards Vapour is irritant to eyes, skin and lungs. It is narcotic. Liquid if swallowed is harmful to liver and kidneys. Highly flammable.
TLV 1000 ppm (2400 mg m^{-3}).
Flashpoint −18°C.
Autoignition temperature 538°C.

Incompatibility Strong oxidising agents, trichloromethane (chloroform) and other halogenated compounds.

Handling Use in well-ventilated area well away from flames, hot plates, etc. Gloves and eye protection advisable.

Storage Flammables store. Up to 250 cm³ in laboratory.

Disposal Wash to waste with plenty of running water.

Spillage Evacuate room if spillage is large. Wear respirator, face shield and gloves; turn off all sources of ignition; wash to waste with water. Ventilate the room thoroughly.

First Aid

Eyes Irrigate with water. Seek medical advice.

Lungs Remove patient from area. Rest and keep warm. Seek medical advice.

Mouth Wash out mouth thoroughly with water. Drink water if liquid has been swallowed. Seek medical advice.

Skin Wash with water.

Local Conditions

P Propyl ethanoate
n-Propyl acetate

$CH_3COOC_3H_7$ Colourless liquid, BP 102°C.

Hazards Liquid is harmful by skin absorption. Vapour is harmful to eyes and lungs. Highly flammable, volatile liquid.
TLV 200 ppm (840 mg m^{-3}).
Flash point 15°C.
Autoignition temperature 450°C.

Incompatibility Strong oxidising agents.

Handling Wear gloves and eye protection. In fume cupboard, away from any source of ignition.

Storage Flammables store. Never in the laboratory.

Disposal Unavoidable discharges, e.g. small quantities such as washings from glassware, etc. should be emulsified and washed to waste.

Spillage Turn off all sources of ignition. Wear gloves and face shield. Small quantities are dealt with as in disposal. Absorb larger quantities on sand and remove to open area for evaporation. Ventilate spillage area well.

First Aid

Eyes Irrigate with water. Seek medical advice.

Lungs Remove patient to fresh air.

Mouth Wash out with water. Seek medical advice.

Skin Wash with soap and water. Wash contaminated clothing.

Local Conditions

C_5H_5N Colourless liquid, BP 115°C.

Hazards   Liquid is harmful to skin and eyes, and very harmful if swallowed. Vapour harmful to eyes and lungs especially if exposure prolonged. Highly flammable, volatile liquid.
TLV 5 ppm (15 mg m⁻³).
Flash point 20°C.
Autoignition temperature 482°C.

Incompatibility   Oxidising agents.

Handling   Wear gloves and eye protection. Handle in fume cupboard, away from any source of ignition.

Storage   Flammables store. Never in laboratory.

Disposal   Small amounts can be washed to waste with plenty of running water.

Spillage   Turn off all sources of ignition. Evacuate laboratory. Wear respirator, face shield and gloves. Mop up with plenty of water and wash to waste. Ventilate spillage area well.

First Aid

Eyes   Irrigate with water. Seek medical advice.

Lungs   Remove patient from exposure, rest and keep warm. Seek medical advice.

Mouth   Wash out thoroughly with water. Seek medical attention.

Skin   Drench with plenty of water. Wash contaminated clothing.

Local Conditions

$C_6H_4NC_3H_3$ Liquid, BP 238°C.

Hazards Liquid is harmful to skin. Vapour is harmful to eyes and lungs. If heated to decomposition it evolves fumes of nitrogen oxides. Prolonged exposure can cause retinitis.
Flash point 99°C.
Autoignition temperature 480°C.

Incompatibility Oxidising agents.

Handling Wear gloves and eye protection.

Storage Flammables store.

Disposal Unavoidable discharges, e.g. small quantities such as washings from glassware, etc. should be emulsified and washed to waste.

Spillage Wear gloves and face shield. Form emulsion with water and detergent. Mop up and wash to waste with large excess of water. Wash site of spillage thoroughly.

First Aid

Eyes Irrigate with water. Seek medical attention.

Lungs Remove patient from exposure, rest and keep warm. Seek medical attention.

Mouth Wash out thoroughly with water. Seek medical attention.

Skin Drench skin with water. Wash contaminated clothing.

Local Conditions

$SiCl_4$ Colourless liquid, BP 59°C. Lachrymatory.

Hazards Corrosive liquid, causes burns. Harmful vapour. Avoid contact with eyes and skin. Avoid breathing vapour. Liquid is poisonous if swallowed. Reacts violently with water to produce toxic fumes of hydrogen chloride.

Incompatibility Water.

Handling Wear gloves and eye protection. Handle in a fume cupboard.

Storage In desiccator over anhydrous calcium chloride as products of hydrolysis are silicon dioxide, which often cements the stopper in position, and hydrogen chloride which creates a high internal pressure. Never in laboratory.

Disposal Wear gloves and face shield. Wash small amounts cautiously to waste in fume cupboard with running water.

Spillage Wear respirator, face shield and gloves. Spread anhydrous sodium carbonate liberally over the spillage, and mop up cautiously with water. Wash to waste using running water.

First Aid

Eyes Irrigate with water. Seek medical attention.

Lungs Remove patient from exposure, rest and keep warm. Seek medical advice.

Mouth Wash out thoroughly with water and give plenty of water to drink, followed by milk of magnesia. Seek medical advice.

Skin Drench with plenty of water. Remove contaminated clothing and wash before re-use. Seek medical advice.

Local Conditions

AgNO$_3$ White solid.

Hazards Corrosive solid, causes burns. Avoid contact with eyes and skin. *TLV* 0.01 mg m^{-3}.

Incompatibility Ethyne (acetylene); ethanedioic acid (oxalic acid); silver nitrate solution with magnesium powder and strong reducing agents; solid with carbon, phosphorus and sulphur. Ammoniacal silver nitrate solution can on standing form explosive azides. Such solutions should be freshly prepared when needed and *immediately* disposed of afterwards by washing to waste with copious amounts of running water.

Handling Wear gloves and eye protection.

Storage General store with oxidising agents. Never in laboratory.

Disposal Wash to waste with running water or into silver residues bottle.

Spillage Wearing gloves and face shield, sweep up and wash to waste with water or into silver residues bottle.

First Aid

Eyes Irrigate with water. Seek medical attention.

Lungs ——

Mouth Wash out mouth thoroughly with water. Seek medical advice.

Skin Drench skin with plenty of water. Remove contaminated clothing and wash before re-use.

Local Conditions

$KAg(CN)_2$ Schedule 1 poison. White solid. This substance should not normally be available in educational laboratories.

Hazards Vapour highly poisonous. Also evolves hydrogen cyanide in contact with moist air. Corrosive to skin.
TLV (skin) 5 mg m^{-3} (as CN).

Incompatibility Water, nitrites (nitrates (III)), acids (even carbon dioxide of the air).

Handling In the fume cupboard. Wear gloves and eye protection.

Storage Poisons cupboard, not in laboratory.

Disposal For large amounts consult Local Authority. For small amounts wear gloves, face shield and respirator, dissolve in water, add excess of sodium chlorate (I) (sodium hypochlorite, household bleach, or bleaching powder) and allow to react for a period of 24 hours. The cyanide is converted to cyanate which can be washed down the drain with running water.

Spillage Evacuate the room. Wear respirator, face shield and gloves. For spillage of solution, scatter bleaching powder on top and mop up leaving the treated cyanide in a bucket for 24 hours before washing to waste. Consult Local Authority if spillage is large. Solid should be carefully brushed up, and treated as in disposal.

See over for First Aid

First Aid

Rescuers must wear respirators whilst giving first aid since there is a risk of vapour inhalation from contaminated clothing.

Obtain medical attention at once.

If casualty is breathing, break capsule of pentyl nitrite (amyl nitrite) and give to inhale for 15–20 seconds. Repeat every 2 or 3 minutes to a maximum of two capsules. Administer oxygen through a face mask, keep warm and rest.

Eyes Irrigate thoroughly with water. Seek medical attention.

Lungs Remove patient from exposure. Remove all clothing and place in open air. If breathing has stopped apply artificial respiration *other than mouth to mouth, or mouth to nose. Obtain medical attention at once.*

Mouth Make sick by giving cyanide antidote or finger down back of throat. *Obtain medical attention at once.*

Skin Wash with water. Remove all clothing, putting it in open air. Seek medical advice.

Local Conditions

Mixture of $NaOH$ and $Ca(OH)_2$ White solid.

Hazards   Corrosive solid. Avoid contact with skin and eyes. Avoid breathing the dust. Gives a strongly corrosive solution.

Incompatibility   Strong acids.

Handling   Wear gloves and eye protection.

Storage   General store with alkalis. Never in laboratory.

Disposal   Wash to waste with plenty of water.

Spillage   Wear gloves and face shield. Shovel into a dry bucket. Add, a little at a time, to excess water and wash to waste, diluting greatly.

First Aid

Eyes   Irrigate with water. Seek medical attention.

Lungs   Remove patient from exposure. Rest and keep warm. Seek medical attention.

Mouth   Wash out mouth with water. Give plenty of water to drink. Seek medical advice.

Skin   Drench with plenty of water. Remove contaminated clothing and wash before re-use.

Local Conditions

Sodium (metal)

Soft silvery solid, usually coated with grey oxide.

Hazards A flammable corrosive solid. Dangerous when exposed to heat or flame, or by chemical reaction with moisture, air or any oxidising material. Spontaneously flammable when heated in air. Reacts violently with water to produce highly flammable hydrogen gas, and a corrosive solution of sodium hydroxide.

Incompatibility Water, acids, carbon dioxide, tetrachloromethane (carbon tetrachloride) and other chlorinated hydrocarbons, halogens and oxidising agents.

Handling Wear gloves and eye protection.

Storage With reducing agents. Store under oil. Never in laboratory.

Disposal Wear gloves and face shield. Add small pieces, no bigger than a pea, one piece at a time to a beaker containing propan-2-ol, in a fume cupboard away from sources of ignition. When the reaction is complete, neutralise and wash to waste with water.

Spillage For small amounts, deal with as in disposal. For large amounts, cover with anhydrous sodium carbonate and shovel into a dry bucket. Transport to a safe, open area and add, a little at a time, to dry propan-2-ol. Leave to stand for 24 hours, neutralise and wash to waste, diluting greatly with running water.

First Aid

Eyes Flood with plenty of water. Seek medical attention.

Lungs ——

Mouth Wash out mouth with water. Seek medical attention.

Skin Drench skin with water after removing any adhering metal. Seek medical advice.

Local Conditions

Grey spongy mass of sodium and mercury stored in liquid paraffin.

Hazards As for sodium and mercury but action with water, etc. is much reduced. Presence of sodium makes it caustic and flammable. Presence of mercury makes it very poisonous.

Incompatibility Water, carbon dioxide, tetrachloromethane and (as for sodium) other chlorinated hydrocarbons, halogens.

Handling Wear gloves and eye protection. Keep away from water.

Storage With reducing agents. Never in laboratory.

Disposal Small amounts can be added to large quantity of water in a beaker or trough. This must be done in a well-ventilated place since hydrogen will be evolved. When sodium has all reacted, the mercury can be recovered and the sodium hydroxide solution neutralised and washed to waste with water.

Spillage Small amounts can be collected on a shovel and treated as for disposal. Spread paste of calcium hydroxide and flowers of sulphur over area of spillage as for mercury. For large amounts, wear gloves and face shield, cover with anhydrous sodium carbonate and shovel into a dry bucket. Transfer to an open area and add a little at a time to a large quantity of water. Leave for 24 hours and finally recover the mercury.

First Aid

Eyes Irrigate with water. Seek medical attention.

Lungs ——

Mouth Wash out with water. Seek medical attention.

Skin Wash thoroughly with water. Seek medical advice.

Local Conditions

$NaBrO_3$ White solid.

Hazards	A powerful oxidising agent. May react violently with oxidisable material, often causing combustion. Harmful if swallowed. More toxic than chlorates (V). Dangerous when heated to decomposition, gives fumes of bromine.
Incompatibility	Reducing agents and any oxidisable material.
Handling	Normal.
Storage	General store with oxidising agents.
Disposal	Wash to waste with water.
Spillage	Sweep up and wash to waste with water. Wash area well to remove bromate which makes organic material (wooden floor, bench, etc.) more combustible.

First Aid

Eyes	Irrigate with water. Seek medical attention.
Lungs	——
Mouth	Wash with water. Seek medical attention.
Skin	Wash thoroughly with water. Seek medical advice.

Local Conditions

NaClO   Colourless solution smelling of chlorine.

Hazards   Corrosive, causes burns. Reacts with acids to evolve chlorine gas. Avoid contact with skin and especially eyes. Dangerous when heated, evolves chlorine.
TLV of chlorine is 1 ppm.

Incompatibility   Acids, many organic compounds, ammonia, reducing agents. Decomposed to sodium chloride and oxygen in presence of suitable catalysts (cobalt and other transition metal compounds).

Handling   Wear gloves and eye protection.

Storage   General store. Up to 500 cm³ in laboratory.

Disposal   Wear gloves and face shield. Wash to waste with plenty of water.

Spillage   Wear gloves and face shield. Mop up and wash to waste with plenty of water.

First Aid

Eyes   Irrigate with water. Seek medical attention.

Lungs   Remove patient from exposure, rest and keep warm. Seek medical advice.

Mouth   Wash out mouth thoroughly with water, give large quantities of water to drink. Seek medical attention.

Skin   Drench with plenty of water. Remove contaminated clothing and wash before re-use. Seek medical advice.

Local Conditions

C_2H_5ONa White solid.

Hazards Corrosive solid, causes burns and affects skin and eyes. Highly flammable.

Incompatibility Acids.

Handling Wear gloves and eye protection. This will rarely be bought and stored, but is prepared in laboratory.

Storage Flammables store. Never in the laboratory.

Disposal Wear gloves and face shield. Add to a large quantity of water. When reaction is complete wash to waste diluting greatly with water.

Spillage Wear gloves and face shield. Shovel into a dry bucket. Remove to a safe area and add, a little at a time, to a large quantity of water. When the reaction is complete wash to waste with water.

First Aid

Eyes Irrigate with water. Seek medical attention.

Lungs ——

Mouth Wash out mouth thoroughly with water. Give plenty to drink. Seek medical attention.

Skin Drench with plenty of water. Remove contaminated clothing and wash before re-use. Seek medical advice.

Local Conditions

NaH   White solid.

Hazards This solid reacts violently with water, releasing highly flammable hydrogen gas and forming sodium hydroxide. Reacts violently with strong oxidising agents. Liberates hydrogen when heated. Sodium hydride dust can explode due to contact with flames, sparks, heat or oxidising agents. Avoid contact with skin and eyes.

Incompatibility Water, oxidising agents, acids.

Handling Wear gloves and eye protection. Handle in a fume cupboard, away from flames and moisture.

Storage General store with reducing agents. Never in laboratory.

Disposal Wear gloves and face shield. In a well-ventilated area, add small quantities slowly to large quantity of water and when reaction complete neutralise and wash to waste.

Spillage Wear gloves and face shield, shovel the solid into a dry bucket. For small amounts sprinkle slowly into a large quantity of water and finally wash to waste. For large amounts, sprinkle on ground in a safe area and spray with water.

First Aid

Eyes Irrigate with water. Seek medical attention.

Lungs Remove patient from exposure. Rest and keep warm.

Mouth Wash thoroughly with water. Give plenty of water to drink. Seek medical attention.

Skin Drench with water. Remove contaminated clothing and wash before re-use. Seek medical advice.

Local Conditions

NaOH   White solid.

Hazards Corrosive solid. Skin contact is harmful. Much heat is evolved when the solid is added to water. The solution can cause severe burns. Very dangerous to eyes. Avoid contact with eyes and skin. (C) *TLV* 2 mg m^{-3} (dust)

Incompatibility Strong acids. Also water (high heat of solution), zinc, aluminium.

Handling Wear gloves and eye protection.

Storage General store. Never in the laboratory. Do not store solid or solutions in ground glass stoppered bottles since these stick very badly.

Disposal Wear gloves and face shield. Dissolve in large volume of water, neutralise and wash to waste with plenty of water.

Spillage Wear gloves and face shield. Shovel into a dry plastic bucket, transport to a safe open area and add, a little at a time, to a large quantity of water. Neutralise and wash to waste with plenty of water.

First Aid

Eyes Irrigate with water. Seek medical attention.

Lungs ——

Mouth Wash thoroughly with water. Give plenty of water to drink. Seek medical attention.

Skin Drench with water. Remove contaminated clothing and wash before re-use. Seek medical advice.

Local Conditions

$NaIO_3$　White solid.

Hazards	Powerful oxidising agent. May cause fire if left in contact with combustible material. Poisonous if swallowed, similar to chlorates (V) and bromates (V).
Incompatibility	Reducing agents.
Handling	Use eye protection.
Storage	With oxidising agents. Not in laboratory.
Disposal	Wearing face shield and gloves, wash to waste with water.
Spillage	Sweep up and wash to waste with water. Wash area well to remove iodate which makes organic material (e.g. wooden floor, bench) more combustible.

First Aid

Eyes	Irrigate with water. Seek medical attention.
Lungs	——
Mouth	Wash out mouth with water. Seek medical attention.
Skin	Wash thoroughly with water. Seek medical advice.

Local Conditions

Sodium nitrite
Sodium nitrate (III)

$NaNO_2$ White or yellowish solid.

Hazards	Strong oxidising agent. Poisonous; as little as 1 g can prove fatal. Inhalation of dust dangerous. Can explode by friction if in contact with organic matter. Explodes when heated to over approx 500°C. Chronic ingestion may cause cancer (Sax).
Incompatibility	With cyanides, ammonium salts, thiosulphates it can form explosive mixtures. With acids, nitrogen oxides are evolved.
Handling	Gloves and eye protection.
Storage	Poisons cupboard.
Disposal	Wash small amounts to waste with plenty of water. If amounts large consult Local Authority
Spillage	Mop up with plenty of water and wash to waste with plenty of water. Wash site of spillage well especially if wood.

First Aid

Eyes	Irrigate with water. Seek medical attention.
Lungs	Seek medical advice.
Mouth	Wash with water. Seek medical attention.
Skin	Wash thoroughly with water. Wash clothing.

Local Conditions

Na_2O White solid.

Hazards Corrosive solid. Dissolves in water to form corrosive solution of sodium hydroxide. Avoid contact with eyes and skin.

Incompatiblity Water, acids.

Handling Wear gloves and eye protection.

Storage General store with alkalis. Never in laboratory.

Disposal Wear gloves and face shield. Add cautiously in small quantities to a large quantity of water, neutralise and wash to waste with plenty of water.

Spillage Wear gloves and face shield. Sweep up into a dry bucket, and then treat as in disposal.

First Aid

Eyes Irrigate with water. Seek medical attention.

Lungs Remove patient from exposure, rest and keep warm.

Mouth Wash the mouth thoroughly with water. Give plenty of water to drink. Seek medical attention.

Skin Drench with water. Remove contaminated clothing and wash before re-use.

Local Conditions

Na_2O_2 White solid.

Hazards   Highly toxic, corrosive solid, causes burns. Powerful oxidising agent. Do not breathe dust. Avoid contact with skin and eyes. Fire hazard – use large quantities of water and not chemical extinguishers.

Incompatibility   Ethanoic acid (acetic acid); ethanoic anhydride (acetic anhydride); benzaldehyde; carbon disulphide; ethyl ethanoate (ethyl acetate); ethane-1,2-diol (ethylene glycol); furfural; propane-1,2,3-triol (glycerol, glycerine); hydrocarbons; methanol; oxidisable materials; powdered metals; water.

Handling   Wear gloves and eye protection.

Storage   With oxidising agents. Never in laboratory.

Disposal   Wear gloves and face shield. Never in waste bin. Add small quantities very cautiously to a large volume of water. When the reaction is complete, neutralise and wash to waste with plenty of water.

Spillage   Wear gloves, face shield and dust respirator. Shovel into a dry bucket, transport to a safe open area, then treat as in disposal. Wash spillage area well.

First Aid

Eyes   Irrigate with water. Seek medical attention.

Lungs   Remove patient from exposure. Rest and keep warm. Seek medical advice.

Mouth   Wash mouth thoroughly with water. Give plenty of water to drink. Seek medical attention.

Skin   Drench with water. Remove contaminated clothing and wash before re-use. Seek medical advice.

Local Conditions

Sr   Solid

Hazards   Harmful by contact, swallowing or inhalation of dust. Latter unlikely since it is stored under liquid paraffin. Flammable.

Incompatibility   Reacts with water to produce hydrogen and heat. Oxidising agents. Very similar in reactions to sodium.

Handling   Wear gloves and eye protection. Remove from liquid paraffin for only a short time when cutting off small pieces. Use at once.

Storage   With reducing agents. Never in laboratory.

Disposal   Wear face shield and gloves. Add very small pieces to propan-2-ol. Leave until reaction is complete and add to large volume of water. Neutralise with hydrochloric acid and wash to waste with plenty of water.

Spillage   Wear gloves and face shield. Cover with anhydrous sodium carbonate and shovel into dry bucket and in safe area add a little at a time to large excess of dry propan-2-ol. Leave for 24 hours, neutralise with hydrochloric acid and wash to waste with plenty of water.

First Aid

Eyes   Flood with plenty of water. Seek medical attention.

Lungs ——

Mouth   Wash with water. Seek medical attention.

Skin   Wash with water.

Local Conditions

Yellow solid.

Hazards   Low toxicity. Fire hazard slight but when burning produces toxic sulphur dioxide.
Autoignition temperature 247°C.

Incompatibility   Dangerous explosive mixtures with many metallic powders (especially zinc and magnesium), oxidising agents such as chlorates (V), ammonium nitrate, nitrates and nitrites of sodium and potassium, chromic (VI) acid, mercury (II) oxide, chlorates (VII) (perchlorates).

Handling   Normal.

Storage   General store.

Disposal   Mix with moist sand, place in bag, seal and put in refuse.

Spillage   As in disposal.

First Aid

Eyes   Irrigate with water. Seek medical advice.

Lungs —

Mouth   Wash with water. If swallowed seek medical advice.

Skin   Wash with water.

Local Conditions

SO_2Cl_2 Pungent colourless liquid, BP 67°C. Turns yellow on standing due to dissociation into chlorine and sulphur dioxide.

Hazards Corrosive liquid and vapour attacks skin, eyes and mucous membranes. Dangerous if swallowed or inhaled. Emits fumes of chlorides and sulphur oxides if heated to decomposition.

Incompatibility Hydrolysed slowly by water to give sulphuric and hydrochloric acids. Reacts violently with alkalis, lead (IV) oxide (lead dioxide), phosphorus and higher oxides of nitrogen.

Handling Wear gloves and eye protection. In fume cupboard.

Storage Acids cupboard.

Disposal Wear face shield and gloves. Using good ventilation or a fume cupboard add slowly to a large volume of water in a plastic bucket and allow to stand for twenty-four hours. When reaction is complete neutralise with sodium carbonate and wash to waste with running water.

Spillage Ventilate and evacuate room. Wearing respirator, gloves and face shield, cover with anhydrous sodium carbonate. Mop up carefully and wash to waste with running water.

First Aid

Eyes Irrigate with water. Seek medical attention.

Lungs Remove patient from area. Seek medical attention.

Mouth Wash out well with water. If swallowed give water to drink. Seek medical attention.

Skin Wash with water. Seek medical advice.

Local Conditions

Sulphur dichloride oxide
Thionyl chloride

$SOCl_2$ Pale yellow liquid, BP 79°C. Lachrymatory.

Hazards   Quite toxic, pungent fumes are evolved from the yellow liquid. The vapour irritates eyes, skin and respiratory system. The liquid burns eyes and skin. If swallowed the liquid causes serious internal damage.

Incompatibility   Water reacts with it to produce hydrogen chloride and sulphur dioxide.

Handling   Wear gloves and eye protection and handle in fume cupboard.

Storage   With acids in general store. Never in laboratory.

Disposal   Wear gloves and face shield. In a fume cupboard, carefully mix with anhydrous sodium carbonate. Add this mixture carefully to a large volume of water, neutralise and wash to waste with running water.

Spillage   Wear gloves, eye protection and respirator. Spread anhydrous sodium carbonate over the liquid, mop up carefully with water and wash to waste with plenty of running water.

First Aid

Eyes   Irrigate with water. Seek medical attention.

Lungs   Remove patient to fresh air, rest and keep warm. Seek medical attention.

Mouth   Wash with water. Drink water followed by milk of magnesia. Seek medical attention.

Skin   Wash with water. Apply magnesia/glycerol paste.

Local Conditions

SO_2 Colourless gas with a distinctive odour, BP $-10°C$. Lachrymatory.

Hazards Concentrations of 6 to 12 ppm irritate the respiratory system and even a small amount may affect persons suffering from bronchitis and asthma.
TLV 5 ppm (13 mg m^{-3}). Irritant to eyes at 20 ppm.

Incompatibility Sodium hydride, potassium chlorate (V), sodium, finely divided chromium.

Handling Wear gloves and eye protection. Experiments producing sulphur dioxide and use of siphon or cylinder should be carried out in fume cupboard.

Storage Sulphur dioxide cylinders are susceptible to corrosion particularly of the valve and in the area where the brass valve is in contact with the aluminium cylinder. They should therefore be stored out of contact with corrosive fumes from acids. A suitable place is an open shelf on the general store. Never in laboratory. In humid conditions, store in polythene bag containing silica gel.

Disposal Wear gloves and face shield. Leaking cylinder should be placed in a fume cupboard. If a valve on a cylinder cannot be opened, the cylinder should be placed in the freezing compartment of a laboratory fridge or deep freeze for at least 4 hours. The pressure in the cylinder is reduced to about atmospheric. It is then quite safe to file the cylinder at the neck until a very small hole is produced. The gas can then be allowed to escape slowly in fume cupboard or in an open area.

Spillage Evacuate room and ventilate well.

First Aid

Eyes Irrigate with water. Seek medical advice.

Lungs Remove from exposure, rest and keep warm. Seek medical attention.

Mouth Wash with water. Seek medical advice.

Skin Wash with water.

Local Conditions

H_2SO_4 Colourless oily liquid, BP 330°C. Concentrated acid is approximately 98% by weight of acid.

Hazards Very corrosive to eyes, skin and many materials. Considerable heat of dilution with water and mixing with water may produce spraying. Dilute acid is also harmful to eyes and skin. *TLV* 1 mg m^{-3}.

Incompatibility Water. Concentrated acid forms dangerous and even explosive mixtures with propanone (acetone), chlorates (V), chlorates (VII) (perchlorates), manganates (VII) (permanganates).

Handling Wear gloves and eye protection. *Dilute* by adding slowly to water in a beaker while stirring. The beaker should be in trough of water nestling in a sink so that any spillage during dilution will be contained. Reagent bottles should not be more than half full.

Storage Acids cupboard. Concentrated acid should not be stored on open shelves in the laboratory.

Disposal Wear gloves and face shield. Dilute, neutralise with sodium carbonate and wash to waste slowly with plenty of running water. Hot concentrated acid left from a preparation should first be cooled.

Spillage Wear gloves and face shield. For large spillage wear rubber boots. Spread sodium carbonate over the spillage, add water. Mop up and wash to waste with plenty of running water. Mop the area of spillage thoroughly with water.

First Aid

Eyes Irrigate with water. Seek medical attention.

Lungs Seek medical advice.

Mouth Wash out mouth with water. If swallowed give plenty of water to drink, followed by milk of magnesia (at least 1 tablespoonful). Seek medical attention.

Skin Wash at once with large volume of water. Remove contaminated clothing. Seek medical advice.

Local Conditions

CCl$_4$ Liquid, BP 77°C.

Hazards   Very toxic. Harmful to eyes, lungs. Also poisonous if swallowed and by skin absorption. Has narcotic action. A suspected carcinogen (Sax). When heated to decomposition gives highly toxic fumes of carbonyl chloride (phosgene). This can happen if tetrachloromethane is used to extinguish a fire or if a cigarette, cigar or pipe is smoked in atmosphere containing tetrachloromethane. An alternative solvent which is less hazardous is 1,1,1-trichloroethane.
TLV (skin) 10 ppm (65 mg m^{-3}).

Incompatibility   Calcium, potassium, sodium, lithium.

Handling   In fume cupboard. Wear gloves, eye protection and respirator if large quantities are being handled.

Storage   Poisons cupboard.

Disposal   Emulsify with water and detergent and wash to waste.

Spillage   Evacuate room. Wearing respirator and gloves, apply detergent, work to an emulsion with brush and water. Alternatively, absorb on sand and shovel into bucket. Remove to open area for evaporation. Spillage area should be thoroughly washed with soap and water.

First Aid

Eyes   Irrigate with water. Seek medical attention.

Lungs   Remove patient from exposure, rest and keep warm. Seek medical advice.

Mouth   Wash out thoroughly with water. Seek medical advice.

Skin   Wash with water, then with soap and water. Remove contaminated clothing and air thoroughly outside.

Local Conditions

Schedule 1 poisons. These compounds should not normally be available in school laboratories.

Hazards Solid and solution harmful to eyes, skin and respiratory system. Poisonous by swallowing and skin absorption. A suspected carcinogen (EH15/77).
TLV (skin) 0.1 mg m^{-3} as Tl.

Incompatibility Compounds when heated produce very toxic fumes.

Handling Wear gloves, eye protection and respirator, and use a fume cupboard.

Storage Poisons cupboard.

Disposal *Very small* quantities of soluble compounds such as washings from glassware can be dissolved in a large volume of water and washed to waste with running water. Insoluble compounds, deal with as described in *Spillage*.

Spillage Mix with sand and shovel into a plastic bucket. Consult Local Authority.

First Aid

Eyes Irrigate with water. Seek medical advice.

Lungs Remove patient to fresh air, rest and keep warm. Seek medical advice.

Mouth Wash with water. If swallowed give plenty of water to drink. Seek medical attention.

Skin Wash with soap and water. Wash contaminated clothing. Seek medical advice.

Local Conditions

Mixture of Fe_2O_3 powder and Al powder.

Hazards Toxicity as for aluminium and iron compounds. Fire hazard is considerable when exposed to heat or flames. The reaction between Fe_2O_3 and Al when started is difficult to stop. Temperature attained is about 2500°C. For demonstration reaction only a very small quantity should be used since there can be considerable scattering of the reacting materials. The mixture is 3 to 1 by mass of iron (III) oxide and aluminium powder both previously dried, and the mixing should be done gently on paper using a plastic or wooden spatula. Any mixing of powdered metals with oxidising agents should be done as above and preferably in a fume cupboard or behind a safety screen.

Incompatibility Combustible materials.

Handling Make up only the amount of mixture that is required for the demonstration as directed above. Wear face shield.

Storage Do not store the mixture, make up when required.

Disposal Wet with water and mix with sand, put in normal refuse bin.

Spillage As for disposal.

First Aid

Eyes Wash with water. Seek medical advice.

Lungs ——

Mouth Wash with water.

Skin Wash well with water.

Local Conditions

NH_2CSNH_2 White crystalline solid, MP 177°C.
Should only be used in advanced work and under supervision.

Hazards   Very harmful if swallowed. Mildly irritating to skin, eyes and respiratory tract. Prolonged exposure to low concentrations dangerous. A suspected carcinogen (Sax). Emits dangerous fumes if heated to decomposition.

Incompatibility   Oxidising agents.

Handling   Wear gloves and eye protection. Avoid raising dust.

Storage   Poisons cupboard.

Disposal   Wear gloves and eye protection. If amounts small wash to waste. If large consult Local Authority.

Spillage   Wear gloves and eye protection. Dampen to reduce likelihood of dust. Sweep up carefully and if amounts small wash to waste. If large consult Local Authority.

First Aid

Eyes   Irrigate with water. Seek medical advice.

Lungs   Seek medical advice.

Mouth   Wash out with water. Seek medical attention.

Skin   Wash well with water.

Local Conditions

$SnCl_4$ Colourless liquid, BP 114°C. Lachrymatory.

Hazards A fuming liquid producing hydrogen chloride in contact with moisture in the air. Fumes and liquid burn eyes and skin and irritate the respiratory system. Severe internal irritation if swallowed.
TLV 2 mg m^{-3} (as Sn).

Incompatibility In contact with water or when heated it reacts vigorously with production of heat and toxic fumes.

Handling Wear gloves and eye protection. In fume cupboard, well away from water.

Storage With acids in general store. Never in laboratory.

Disposal Wear gloves and face shield. In fume cupboard, sprinkle into large volume of water, neutralise and wash to waste with water.

Spillage Wear gloves, face shield and respirator. Spread anhydrous sodium carbonate on the liquid and mop up cautiously with water. Wash to waste with plenty of running water.

First Aid

Eyes Irrigate with water. Seek medical attention.

Lungs Remove patient from area. Rest and keep warm. Seek medical advice.

Mouth Wash with water, drink water followed by milk of magnesia. Seek medical attention.

Skin Wash with water and apply magnesia/glycerine paste. Wash contaminated clothing.

Local Conditions

T 1,1,1-Trichloroethane
Methyl chloroform

CH_3CCl_3 Colourless liquid, BP 74°C.

Hazards The vapour is harmful to skin, eyes and lungs. Harmful if swallowed. Thought to be less harmful than either trichloroethene or tetrachloromethane. Only flammable with high concentrations of vapour at high temperatures.
TLV 350 ppm (1900 mg m^{-3}).

Incompatibility Powerful oxidising agents, alkali and other metals.

Handling Wear gloves and eye protection and either work in fume cupboard or in well-ventilated area.

Storage General store. Not in laboratory.

Disposal Wear eye protection and gloves. Unavoidable discharges, e.g. small quantities such as washings from glassware, etc. should be emulsified and washed to waste.

Spillage Wearing gloves and face shield, emulsify with water and wash to waste. Alternatively absorb on sand and remove to outside for evaporation.

First Aid

Eyes Irrigate with water. Seek medical advice.

Lungs Remove patient to fresh air, rest and keep warm. Seek medical advice.

Mouth Wash with water. Seek medical advice.

Skin Wash with soap and water.

Local Conditions

$CCl_3CH(OH)_2$ White solid.

Hazards	Poisonous, corrosive crystals; liquid and vapour irritant to eyes, lungs and skin; slight fire hazard when heated.
Incompatibility	Oxidising agents.
Handling	Wear gloves and eye protection and use in fume cupboard or well-ventilated area.
Storage	Poisons cupboard.
Disposal	Consult Local Authority.
Spillage	Wear gloves and eye protection. Sweep into suitable container and consult Local Authority for disposal.

First Aid

Eyes	Irrigate thoroughly with water. Seek medical attention.
Lungs	Remove patient from exposure, rest and keep warm. Seek medical attention.
Mouth	Wash out thoroughly with water. Seek medical attention.
Skin	Wash thoroughly with water, then with soap and water. Remove contaminated clothing and wash. Seek medical advice.

Local Conditions

T Trichloroethene
Trichloroethylene

$CHClCCl_2$ Colourless liquid, BP 87°C.

Hazards Liquid has sweetish chloroform odour. Vapour is harmful to lungs, skin and eyes. Inhalation of the vapour or ingestion of the liquid may cause headache, nausea and disorientation. Is flammable only with high concentration of vapour in air at high temperatures.
TLV (skin) 100 ppm (535 mg m^{-3}).
Autoignition temperature 410°C.

Incompatibility Powerful oxidising agents; can explode with alkali metals, alkaline earth metals, and with aluminium.

Handling Wear gloves and eye protection. In well-ventilated area.

Storage General store. Never in laboratory.

Disposal Wear gloves and face shield. Unavoidable discharges, e.g. small quantities such as washings from glassware, etc. should be emulsified and washed to waste.

Spillage Wear gloves and face shield. Emulsify with water and detergent and wash to waste with running water. Alternatively, absorb on sand and remove to outside for evaporation.

First Aid

Eyes Irrigate with water. Seek medical attention.

Lungs Remove patient to fresh air, rest and keep warm. Seek medical attention.

Mouth Wash with water. Seek medical attention.

Skin Wash with soap and water. Seek medical advice.

Local Conditions

$CHCl_3$ Liquid, BP 61°C.

Hazards   Vapour is irritant to eyes, skin and lungs. The vapour is an anaesthetic and causes headache, nausea, sickness and unconsciousness. Prolonged inhalation leads to cardiac failure and death. Dilation of pupils is indication of fair degree of exposure. Liquid if swallowed is poisonous. Non-flammable but like tetrachloromethane (carbon tetrachloride) may form carbonyl chloride (phosgene) at high temperature in air due to oxidation. A suspected carcinogen (Sax and EH15/77).
TLV   25 ppm (120 mg m^{-3}).

Incompatibility   Oxidising agents, alkaline propanone (acetone), alkali metals, alkalis.

Handling   Use in well-ventilated area, preferably fume cupboard, and avoid breathing near high vapour concentration. Gloves and eye protection advisable.

Storage   Poisons cupboard.

Disposal   Unavoidable discharges, e.g. small quantities such as washings from glassware, etc. should be emulsified and washed to waste.

Spillage   Wear respirator and face shield. Evacuate room. Emulsify with water and detergent, mop up and wash to waste with running water. Maximum ventilation is essential and room should not be occupied until all trace of the trichloromethane (chloroform) vapour has gone. A spillage can also be absorbed on sand and taken to an open area for evaporation.

First Aid

Eyes   Irrigate with water. Seek medical attention.

Lungs   Remove patient from exposure, rest and keep warm. Seek medical attention.

Mouth   If liquid is swallowed wash out the mouth with water. Seek medical attention.

Skin   Wash with soap and water.

Local Conditions

Triiodomethane
Iodoform

CHI_3 Yellow crystals, MP 120°C.

Hazards Harmful by inhaling, ingesting, skin contact. Heating to decomposition emits fumes of iodine and iodine compounds.
TLV 0.2 ppm (3 mg m^{-3}).

Incompatibility Oxidising agents and alkali metals.

Handling Use gloves and eye protection; handle in fume cupboard.

Storage General store.

Disposal Dissolve small quantities in methylated spirits and emulsify with detergent and water. Wash to waste.

Spillage Shovel into waste bucket after wetting with water. Wash spillage area with detergent and water.

First Aid

Eyes Irrigate with water. Seek medical advice.

Lungs Seek medical advice.

Mouth Wash out with water. Seek medical advice.

Skin Wash with soap and water until yellow stain removed.

Local Conditions

$(NO_2)_3C_6H_2(OH)$ Yellow crystals, must be kept under water.

Hazards Poisonous, corrosive solid which can be absorbed through the skin. The dry solid is a powerful explosive.
TLV (skin) 0.1 mg m^{-3}.
Flash point 150°C.
Autoignition temperature 300°C.

Incompatibility Copper, lead, zinc, silver, ammonia, metal oxides. Many of its salts can explode by friction.

Handling Wear gloves and eye protection. Keep moist and do not heat.

Storage General store. Keep moist with not less than half its own weight of water.

Disposal Dissolve small quantities in hot water and wash to waste with plenty of water.

Spillage Wearing gloves and face shield, moisten well with water and shovel into plastic bucket. For small quantities, add hot water to dissolve and wash to waste with running water.

First Aid

Eyes Irrigate with water. Seek medical attention.

Lungs Remove from area. Seek medical attention.

Mouth Wash with water. If any acid is swallowed give a large quantity of milk to drink. Seek medical attention.

Skin Drench with plenty of water and wash with soap and water. Wash contaminated clothing. Seek medical advice.

Local Conditions

Trioxygen
Ozone

O_3 Gas, BP $-112°C$.

Hazards   This gas is not usually prepared but may be formed during the use of ultra-violet lamps, photocopiers, induction coils, etc. Concentrations as low as 1 ppm have a disagreeable odour and may cause headaches and irritation of the upper respiratory system. *TLV* 0.1 ppm (0.2 mg m^{-3}).

Incompatibility   Reducing agents and organic compounds.

Handling   Any apparatus producing ozone should only be used in a well-ventilated area.

Storage   ——

Disposal   Ventilate.

Spillage   ——

First Aid

Eyes   Remove patient from area. Irrigate eyes with water.

Lungs   Remove patient from area. If exposure serious seek medical advice.

Mouth   ——

Skin   ——

Local Conditions

Mainly $C_{10}H_{16}$ BP 154–170°C.

Hazards Quite harmful by inhalation, swallowing, skin absorption. An allergen. Can cause serious injury to kidneys. Flammable, emits acrid fumes when heated. Dusters, etc. soaked in it are subject to spontaneous combusion and should be dampened and placed in outdoor bin.
TLV 100 ppm (560 mg m^{-3}).
Flash point 35°C.
Autoignition temperature 293°C.

Incompatibility Oxidising agents.

Handling Wear gloves and eye protection. Use in well-ventilated area away from flames and heaters.

Storage Flammables store. Never in laboratory.

Disposal Wear gloves and eye protection. Unavoidable discharges, e.g. small quantities such as washings from glassware, etc. should be emulsified and washed to waste.

Spillage As for disposal.

First Aid

Eyes Irrigate with water. Seek medical advice.

Lungs Remove patient from area, rest and keep warm. Seek medical advice.

Mouth Wash out. Seek medical attention.

Skin Wash with soap and cold water. Seek medical advice.

Local Conditions

V Vanadium compounds

Hazards	Dust is harmful to eyes, lungs and skin. Very harmful if swallowed. (C) *TLV* 0.5 mg m^{-3} for vanadium (V) oxide dust; 0.05 mg m^{-3} for fumes.
Handling	Wear gloves and eye protection. Avoid producing dust. Avoid contact with skin.
Storage	General store. Never in laboratory.
Disposal	Wear gloves and eye protection. Dissolve soluble compounds in water and wash to waste with running water. Insoluble compounds should be mixed with sand and the Local Authority consulted on disposal.
Spillage	As for disposal

First Aid

Eyes	Irrigate with water. Seek medical advice.
Lungs	Remove patient to fresh air, rest and keep warm.
Mouth	Wash with water. Seek medical attention.
Skin	Wash with soap and water.

Local Conditions

Hazards	Low toxicity. Fumes of zinc oxide from heated or burning zinc harmful but not cumulative. *TLV* of oxide fumes 5 mg m^{-3}. Some hazard in form of dust when exposed to flames. Zinc chloride (often used as a flux in soldering) is corrosive and its fumes can damage skin, respiratory tract and intestines. *TLV* of zinc chloride fumes 1 mg m^{-3}.
Incompatibility	Acids, alkalis. Forms explosive mixtures with sulphur or other strong oxidising agents.
Handling	Avoid raising dust.
Storage	Metal with reducing agents, chloride with acids.
Disposal	Metal powder: wet thoroughly and put in normal waste bin.
Spillage	Wet and shovel into waste bin.

First Aid

Eyes	Wash with water. Seek medical advice.
Lungs	Remove patient from area of contamination, rest and keep warm. Seek medical advice.
Mouth	Wash with water. Seek medical advice.
Skin	Wash with soap and water.

Local Conditions

Bibliography

The Care, Handling and Disposal of Dangerous Chemicals, P. J. Gaston, 1970, Northern Publishers
Code of Practice for Chemical Laboratories, 1976, Royal Institute of Chemistry/Chemical Society
Dangerous Properties of Industrial Materials, Irving Sax, 1975, Van Nostrand Reinhold
Handbook of Reactive Chemical Hazards, L. Bretherick, 1975, Butterworth
Hazards in the Chemical Laboratory, G. D. Muir, 1977, Royal Institiue of Chemistry/Chemical Society
Safeguards in the School Laboratory, ASE, 1976, J. Murray
Safety in Science Laboratories, DES Safety Series No 2, 1978, HMSO
'Safety in Schools – General Aspects' by G. P. D. Gordon and 'Safety in Science Laboratories' by A. W. Jeffrey in Aberdeen College of Education *Biology Newsletter No. 29, 1977*

Statutory Instruments and Circulars

Health and Safety at Work Act, 1974, HMSO
Control of Pollution Act, 1974, HMSO
Deposit of Poisonous Waste Act and Statutory Instrument No. 1017 *The Deposit of Poisonous Waste (Notification of Removal or Deposit) Regulations*, 1972, HMSO
Highly Flammable Liquids and Liquefied Petroleum Gases and *Guide to 1972 regulations with 'Certificates of Approval', etc.* HMSO
The Poison Rules, 1978, HMSO
The Medicines and Poisons Guide, 1978, Pharmaceutical Press
Petroleum (Consolidation) Act, 1928, HMSO
Protection of Eyes Regulations, 1974, HMSO
Carcinogenic Aromatic Amines in Schools and other Educational Establishments, DES AM 3/70 (England and Wales), HMSO
The Use of Carcinogenic Substances in Educational Establishments, SED Circular 825 (Scotland), 1972, HMSO
The Use of Inflammable Materials in Homecraft Classes, SED Memorandum 3/65 (Scotland)
The Use of Asbestos in Educational Establishments, SED Circular 668 (Scotland), 1968, HMSO (An updated version of this document is to be issued.)
The Use of Asbestos in Educational Establishments, DES AM 7/76 (England), AM 5/76 (Wales), HMSO
The Packaging and Labelling of Dangerous Substances Regulations, Health and Safety Executive, 1978, HMSO
The Hazards of Experimenting with Explosives, SED Memorandum 39/66
High Pressure Oxygen (Manifold) Systems, SED Memorandum 8/74
Recommendations for Health and Safety in Workshops of Schools and Colleges, BS 4163, British Standards Institution
Technical Data Notes and Guidance Notes, each dealing with a particular substance or with recommended procedures for handling, storing and maintaining equipment and materials, are available from the Health and Safety Executive.